T0332935

Introduction to the Maths and Physics of Quantum Mechanics

Introduction to the Maths and Physics of Quantum Mechanics details the mathematics and physics that are needed to learn the principles of quantum mechanics.

It provides an accessible treatment of how to use quantum mechanics and why it is so successful in explaining natural phenomena. This book clarifies various aspects of quantum physics such as 'why quantum mechanics equations contain "I", the imaginary number?', 'Is it possible to make a transition from classical mechanics to quantum physics without using postulates?' and 'What is the origin of the uncertainty principle?'. A significant proportion of discussion is dedicated to the issue of why the wave function must be complex to properly describe our "real" world.

The book also addresses the different formulations of quantum mechanics. A relatively simple introductory treatment is given for the "standard" Heisenberg matrix formulation and Schrodinger wave-function formulation and Feynman path integrals and second quantization are then discussed.

This book will appeal to first- and second-year university students in physics, mathematics, engineering and other sciences studying quantum mechanics who will find material and clarifications not easily found in other textbooks. It will also appeal to self-taught readers with a genuine interest in modern physics who are willing to examine the mathematics and physics in a simple but rigorous way.

Key Features:
- Written in an engaging and approachable manner, with fully explained mathematics and physics concepts.
- Suitable as a companion to all introductory quantum mechanics textbooks
- Accessible to a general audience

Lucio Piccirillo is a Professor of Radio Astronomy Technology at the University of Manchester, UK, with extensive experience in designing and building cryogenic systems primarily used to cool astrophysical detectors. He has written more than 100 publications in international journals. He has written more than 100 publications in international journals and is the author of Miniature Sorption Coolers: Theory and Applications (CRC Press, 2018) and Introduction to the Maths and Physics of the Solar System (CRC Press, 2020).

Introduction to the Maths and Physics of Quantum Mechanics

Lucio Piccirillo

CRC Press
Taylor & Francis Group
Boca Raton London New York

CRC Press is an imprint of the
Taylor & Francis Group, an **informa** business

Designed cover image: Shutterstock_582084685

First edition published 2024
by CRC Press
6000 Broken Sound Parkway NW, Suite 300, Boca Raton, FL 33487-2742

and by CRC Press
4 Park Square, Milton Park, Abingdon, Oxon, OX14 4RN

CRC Press is an imprint of Taylor & Francis Group, LLC

ISBN: 978-0-367-70302-8 (hbk)
ISBN: 978-0-367-69901-7 (pbk)
ISBN: 978-1-003-14556-1 (ebk)

DOI: 10.1201/9781003145561

Typeset in Nimbus font
by KnowledgeWorks Global Ltd.

Publisher's note: This book has been prepared from camera-ready copy provided by the authors.

Dedication

To my kids: Emma, Julia, Madelaine, Jacopo and Tommaso:
my life is dedicated to them.
To my Father, my Mother, zio Peppino, zio Mario and zia Loreta:
the pillars of my existence.

Contents

SECTION I Review of Classical Mechanics

SECTION II Quantum Mechanics

Foreword

There are many excellent books that treat the mathematical and physical foundation of quantum mechanics. The question is: why another one? Approaching physics subjects usually progresses like this: a young individual or a curious adult is interested in physics and tries to understand it better by reading a popular science book. By reading it, he/she gets a gist of the subjects – for example, black holes or lasers – but more than clarifying, the book sparks many questions. The most common questions are "why?" or "how do they know?" and so on. There is therefore a moment when he/she is not anymore happy with just a qualitative description of nature but wants to understand better how scientists arrive to often very bold assertions. In practice, this corresponds to understand the way scientists make predictions. Given some amount of knowledge about a physical system, he/she wants to learn how to calculate new things about the system. For example, given position and velocity of a point-like particle of mass m at the time t_0, where will the particle be at the time $t_1 > t_0$?

If the individual in question decides to understand physics systematically, he/she usually attends a University study program where somebody gives lectures and recommends books. Some of the most important initial concepts are: how to measure things and how to express the measurements with numbers with units. The next step is to try to see if there are patterns in the numbers when we try to change something. Suppose the point-like mass is in the gravitational field of the Earth, and we drop it from some height and we measure the position of the mass with time. If we plot the position versus time, we see a beautiful curve emerging from the data: a parabola. It seems therefore that we can use maths to describe the physical world. Nature seems to obey laws that can be expressed in mathematical form. In fact, we can fit the data collected for the falling object with a simple expression $s(t) = s(0) - gt^2$, which allows us to make predictions; i.e. we can predict where how the object will accelerate if we drop it from higher and higher altitudes... or not? But the magic seems to be in the exponent 2 in the previous expression and in the constant g. Why 2? Is it *exactly* 2 or perhaps 1.99998 or 2.000001? Is the exponent 2 due to gravity? If yes, why and how? The regularity or patterns in the data suggests that there is a mathematical relation suggesting, in turn, that there must be some sort of **physical law**. A physical law is a recipe that allows anybody to make predictions given some knowledge of nature: given two systems in exactly the same state, they will evolve with time in exactly the same way. This is more or less classical mechanics with its infinite precision in the knowledge of physical quantities and absolute certainty in the time evolution of a system. Classical mechanics is the common sense of everyday life of human beings when speed is not too high, and distances are not too big or too small.

When nature started to be probed at very small scales – for example, atomic lengths – new phenomena emerged that could not be described with the mathematics used by classical mechanics. New equations and new tools needed to be employed to try to make predictions. However, even the act of measurements needed to be re-interpreted with the introduction of probability amplitudes. In the path followed by students to understand modern physics, quantum mechanics is certainly posing a big obstacle: students need to change the way they think about nature. Objects can be in more than one place at the same time, repeating experiments under the exact same initial conditions can have different results, and material objects seem to have wave-like properties and vice versa.

Each quantum mechanics book has its own way of accompanying the students through the difficult transition from classical mechanics to quantum physics and each student is different. Concepts that are grasped easily by a student might represent an insurmountable obstacle to another. That's where the variety of books is extremely useful. By consulting more than one book, there is a high probability that one specific book has the explanation that finally generates the spark in the eyes of the student that each teacher is very familiar with. The spark that means "I got it!".

This book is written with this exact objective: help students to grasp concepts by presenting the subjects, where possible, in slightly different ways, often more than once. If just one student will claim "I got it" after reading this book, then all the efforts in writing are fully justified.

Preface

Quantum mechanics is the foundation of modern physics, and it is one of the most important and influential scientific theories of the twentieth century. It is the theory that explains the behavior of matter and energy on the atomic and subatomic scales. It has revolutionized our understanding of the natural world and has opened up a new realm of possibilities for technological progress.

This book is an introduction to the maths and physics of quantum mechanics. The goal is to provide an accessible yet comprehensive overview of the theory, its basic principles, and its implications for our understanding of the physical world.

The book begins with an introduction to the maths and physics of classical mechanics, putting the accent on the mathematical formalism that will be later used in quantum mechanics. It then provides an overview of the experiments that were not explained by classical mechanics.

Finally, the book provides a nonrigorous overview of the maths needed to formulate quantum mechanics together with those experiments that have been crucial in confirming the validity of the theory.

This book is intended to serve as a comprehensive introduction to quantum mechanics for readers who have a basic understanding of physics but have never formally studied the subject. It provides an accessible yet comprehensive overview of the theory, its basic principles, and its implications for our understanding of the physical world.

Lucio Piccirillo,
Manchester, United Kingdom, April 2023

Section I

Review of Classical Mechanics

1 Classical Physics

Albert Einstein's famous quote "The most incomprehensible thing about the world is that it is comprehensible", hints at the fact that the Universe seems to be ruled by laws expressed as mathematical expressions. This is not at all obvious: as human beings, we are very much capable of describing natural phenomena using also other means like, for example, poetry or music or other forms of arts. We all have felt a sense of wonder at least once in our lifetime when observing the night sky with its majestic display of luminous objects immersed in a vast space, perhaps of infinite extension. It is then natural to try to understand why we have these point-like objects in the sky emitting a feeble light. If we have time to spare, we notice regularities in the motion of these lights in the sky: they seem to rotate around a fixed point in the sky toward the North. Understanding why they move in such a way is difficult, but perhaps we might understand how they move. There is a huge difference in answering "why" rather than "how" and this book is mostly concerned about "how" things behave in a certain way instead of another.

In our frantic modern life, the moments of awe in front of the stars are rare. Our modern life does not give us the time to stop and think about things that are beyond our close surroundings. Perhaps this was not the same for the ancient people like, for example, the Greek philosophers. Due to a different structure of society, there were quite a number of smart and intelligent individuals that had the time to ponder about the nature and its phenomena. Regularities in the motion of stars (including the Sun) were perhaps the first thing that was noticed, but we ascribe to the Greeks the attempt at investigating more in detail the phenomena. The Greeks started asking "why" certain phenomena happen, and they developed a complex model of reality where natural phenomena were generated by a certain number of arbitrary acts of gods. Earthquakes, for example, were thought to be caused by the anger of the god Poseidon. Similarly, lightnings were caused by the anger of Zeus. Invoking gods answered one of the main questions: "why certain phenomena happens"? And the simple answer was that things happen because it is the gods will. This approach was clearly limiting the way in which people can make predictions about natural phenomena. If natural phenomena happen because at a certain place and at a certain time a certain god decided to make it happen then it is really impossible to make a reliable prediction. The only option left was to elaborate a series of actions to try to "placate" the deities in an attempt at trying to predict future phenomena. If not predict, at least have some control.

The main revolution happened when a group of Greek philosophers abandoned searching for the "why" and shifted their attention to the "how" certain phenomena happened without invoking the act of gods. This represented a major shift in investigating natural phenomena. This shift was due to a group of philosophers from the ancient city of Miletus, whose ruins are located near the modern village of Balat in Aydin Province of Turkey. Thales (c. 624 BC to c. 568 BC) was probably the

DOI: 10.1201/9781003145561-1

most representative of the Greek philosophers that advocated to break from the use of mythology to explain natural phenomena and instead explain the world through a number of reasonable hypotheses. These hypotheses are then subject to open evaluation and criticisms from other philosophers until a sort of consensus is reached. This process is not very different from the modern "scientific" approach. Thales was highly regarded by his contemporaries and in fact he was considered one of the seven sages, together with Pittacus, Bias, Solon, Chilon and a couple more not clearly identified. It is important to notice that this new approach shifted the main questions from "why" to "how". Once the Greeks disregarded the actions of gods, unpredictable by definition, they started to concentrate on finding how things happen. Is nature forced to follow certain rules? The search for these rules, or natural laws, started then and it is still going on today.

Modern science is still attempting to understand the working of the Universe by trying to identify a set of basic "laws" that need to be discovered, usually by looking at the results of experiments or observations. As a result of these observations over thousands of years, we are today in a position to state that natural phenomena are subject to a set of laws. We call these laws "physical laws". These laws derive from careful observations of patterns in natural phenomena and are expressed as "mathematical" relationships. It is extremely important to understand that laws are given without proof and are considered valid until an experiment (or, better, more than one single experiment) shows a violation. In this case, we have a few choices: we try to see if we can modify our laws to include the new experimental data in the range of validity of the law; we scrap the law completely and look for a new one; we re-define the "range of applicability" of the laws under the assumption that other laws are needed outside such a range. The laws don't usually provide an explanation for a mechanism producing such phenomena.

A "Universal Law" is a law that is assumed to be valid everywhere in the Universe, in the past, present and future. This is undoubtedly the strongest statement about the Universe that we can make. We obviously have no way, even theoretically, to verify this kind of law. Let's think about the possibility that the Universe is infinite in spatial extension: any experiment trying to verify **directly** this assumption is destined to run forever...

In addition to laws and universal laws, often in the scientific literature, we encounter the terms *axiom*, *postulate* and *principle*. Axiom and postulates are terms used mostly in mathematics and are referred to a proposition that is assumed to be true to study the consequences that follow from them. Axioms and postulates are, by definition, unprovable and considered self-evident[1]. A principle is a statement about nature, coming from observations, that is considered to be true until disproved experimentally. Laws are expressed mathematically while principles are usually general statements. One example of principles is the statement of constant speed of light irrespective of the relative speed of observers measuring it. One example of laws is Newton's universal law gravitation.

---

[1] In general terms, it is customary to use axioms when dealing with real numbers while postulates are used mostly in geometry.

Before delving into the machinery of classical and quantum physics, let us state what is the basic problem: we want to be able to predict how a system evolves in time given a set of measurable quantities and a reasonable set of assumptions. To be able to describe the evolution of a system, we need to define it first and then determine the laws that govern its time evolution. Classical mechanics and quantum mechanics have dramatically different laws, and this book will try to clarify the differences (and the similarities).

Very often a law is expressed through an equation. In the case of the motion of a body, or a particle, an equation is set up such that the time evolution of a parameter is equated to something known. In this way, by solving the equation, we can "predict" how the parameter we interested "evolves" with time. The case of the trajectory of a particle requires that we can predict the future position of the particle with time. We will see in section 1.1 that Newton's second law is an example of such an equation in which we can predict the future (and past) position of a body of mass m if we know the force F acting on it together with the position and velocity of the body at time $t = 0$.

1.1 CLASSICAL MECHANICS

In this section, we will review the classical mechanics, starting from Newtonian mechanics. We will then progress to Lagrangian and Hamiltonian formulation to finish with Poisson brackets. During this classical excursion, most of the mathematics needed (but not all!) in Quantum Mechanics will be introduced.

We begin our discussion of classical mechanics by discussing what is meant by the term **classical state of a system**. We added the term "classical" to distinguish it from the quantum state of a system which will be discussed later in the book. In rough terms, the state of a system is the collection of information that is needed to instruct somebody to make an identical copy of it. The system can be a single particle or a complex collection of particles including the case of a solid body[2].

In the simple case of a single particle, its classical state at a given time t is completely determined when we give its position and its momentum, respectively $x(t)$ and $p(t)$. If in addition, we want to predict how the system evolves with time, then we need to give the laws of motion, i.e. the equations that predict how the particle evolves its position $x(t)$ and its momentum $p(t)$. More specifically, given an arbitrary (small) time interval dt, the equations of motion give us $x(t + dt)$, $p(t + dt)$ given $x(t)$ and $p(t)$. We will see below that Newton has provided us with a second order ordinary differential equation (ODE) capable of such predictive power.

There is an interesting way to represent the state of a single particle by introducing the **phase space**, i.e. a Cartesian plane where the x-axis is $x(t)$ and the y-axis is $p(t)$. In this space, the state of a particle is a single a point and the equations of motion tell us how this point moves with time. The motion of the point in the phase space defines a phase trajectory. We will see examples of such trajectories later on.

---

[2]In thermodynamics the state of a system is its condition at a specific time, that is fully identified by the values of a set of parameters.

1.1.1 THE POSTULATES OF CLASSICAL MECHANICS

It is the general understanding among physicists that the postulates of Newtonian mechanics are given by the three Newton's laws, given in the next section. We want to point out that there is a difference between "physical laws" and "physical postulates". Postulates are mathematical statements or assumptions that form the theoretical foundation of the laws describing the physical phenomena to be investigated. Postulates are not usually experimentally verifiable. Laws instead are mathematical statements that are experimentally verifiable and are based on the postulates.

We need to be sure that the systems that we are trying to describe are within the range of applicability of the laws. There are two main limitations to be taken into account: high speeds and large densities. In what follows, we will consider that the typical speed of our particle is negligible with respect to the speed of light[3] and that there are no strong gravitational fields[4] usually associated with extremely high mass densities. There is another very important limitation that has a lot to do with this book. In addition to high mass densities and/or large masses, we must also state that classical mechanics is also not applicable when the masses are very small. We will see later in the book that when the masses are very small (compared to our everyday experience), then classical mechanics is not valid, as well.

When we deal with everyday objects with reasonable masses moving at reasonable speed, then classical mechanics is a very good approximation. For several hundred years since Newton all the experiments have had access only to such classical world, i.e. a world of low speed and low mass (but not too low).

It is interesting, and instructive, to review how Newton himself has laid down the foundation of classical mechanics in his book Principia [31]. The Principia, whose full title is "Philosophiae Naturalis Principia Mathematica" translated into "The Mathematical Principles of Natural Philosophy", is organized into three books by Isaac Newton and published in 1687. The first two books deal mainly with the motion of bodies with and without resistance, while the third book concerns mainly to the law of universal gravitation and its consequences to the motions observed in the solar system.

Following Shankar [40] we will now state the *postulates of classical mechanics* for a single particle. Remember that we accept these statements based on our experience of the physical world under the condition that we are ready to abandon whichever postulate is shown to be not in agreement with experimental verification.

I The state of a particle at a time t is completely defined by the two variables $x(t)$ and $p(t)$ which are, respectively, the position and momentum of the particle at time t.

---

[3]The speed of light c is equal to 299,792,458 m/s. More accurately we can still use classical mechanics when the typical speeds involved satisfy the inequality $(v/c)^2 \ll 1$.

[4]We can still use Newton's gravity if $\phi/c^2 \ll 1$ where $G = 6.674 \times 10^{-11} \, m^3 \cdot kg^{-1} \cdot s^{-2}$ is Newton's gravitational constant, $\phi = Gm/r$ is the gravitational potential of a body of mass m and c is the speed of light.

II Every dynamical variable R is only a function of $x(t)$ and $p(t)$.

III If the particle is in a state given by $x(t)$ and $p(t)$, any measurement of the variable R will yield a value $R(x, p)$ and the state is unaltered.

IV The state variables change with time according to the following equations:

$$\dot{x} = \frac{dx}{dt} = \frac{p}{m}$$
$$\dot{p} = \frac{dp}{dt} = F \tag{1.1}$$

1.1.2 NEWTONIAN MECHANICS

We have seen in the previous section that classical mechanics describes the motion of bodies of masses in our range of experience, i.e. from nano-grams to earth/solar masses. We can extend the mass range to the mass of galaxies and cluster of galaxies if we allow general relativity to be considered as a classical theory[5]. In the following, unless otherwise specified, we will be dealing with an idealization of a finite body by introducing the concept of **point mass**. A point mass is a body with a finite mass all concentrated inside a negligible dimension. We will consider this as the definition of a classical **particle**.

We also assume that particles move at speeds much less than the speed of light. We assume that there is a fixed reference frame identified by the distant stars which are supposed to be at rest relative to the absolute space. We call this reference frame as **Inertial Reference Frame**. Time flows identically for all the observers, irrespective of their state of motion; i.e. time is universal. Newtonian mechanics, therefore, is based on the concept of "absolute" space and "absolute" time. One important concept is some time not underlined enough: in classical mechanics we can measure the position and velocity of particles to any accuracy we want and the act of measurement does not affect the results of the measurements. For example, if we want to locate accurately a particle, we can imagine that we shine very intense light upon it of wavelength shorter and shorter the more accurate the position we want. By doing so, we can determine the trajectory and the associated velocities to the accuracy that we want.

Newtonian mechanics is based on three laws, called Newton's laws. As we have mentioned above, a physical law is given without proof and is based on observations of natural phenomena. The same can be stated for Newton's laws. It is worth to mention that in the following, we will consider only the motion of particles with no internal structure. If we want to take into account the internal structure of a body, we need to use a more general set of laws (Euler's laws of motion).

Let's begin with Newton's three laws of motion. For completeness and to give Newton full credit, let us first report the laws as they were written (in Latin) in his book Philosophiae Naturalis Principia Mathematica [31].

---

[5]The straight application of Newtonian mechanics to the motion of stars in most galaxies leads to the necessity of accepting that the majority of the mass is not visible (dark matter).

Lex I: Corpus omne perseverare in statu suo quiescendi vel movendi uniformiter in directum, nisi quatenus illud a viribus impressis cogitur statum suum mutare.
Lex II: Mutationem motus proportionalem esse vi motrici impress, and fieri secundum lineam rectam qua vis illa imprimitur.
Lex III: Actioni contrariam semper and qualem esse reactionem: sive corporum duorum actiones in se mutuo semper esse quales and in partes contrarias dirigi.

A modern translation of the three laws above is given usually in many equivalent different statements. We give here an example:

First Law: In an inertial reference frame a body will either remains at rest or continues to move at a constant velocity unless a force is acted upon it[6].
Second Law: The vector sum of all forces acting on a body is equal to the mass multiplied by the acceleration.
Third Law: To every action there is always a reaction equal and opposite. When one body exerts a force on a second body, the second body exerts a force equal in magnitude and opposite in direction on the first body

The second law is a ordinary differential equation relating the force F acting on a body to the product of its mass m and its acceleration a. The second law is usually written as:

$$F = ma \tag{1.2}$$

A more accurate way to write equation 1.2 would be:

$$F = \frac{dp}{dt} \tag{1.3}$$

where $p = mv$ is the momentum of the body equal to the product of the mass and the velocity (see below). In fig. 1.1 a particle of mass m is moving along the trajectory s. At a certain time t_1 the particle is at the point P identified by the position vector r_1. Under the influence of some force F, the particle will move to another point identified by the position vector r_2. We indicate with Δs the displacement vector such that $r_1 + \Delta s = r_2$. Notice that Δs is an approximation to the curved trajectory followed by the particle. Taking shorter and shorter Δs, i.e. having r_2 approaching closer and closer, r_1 approximates better and better the "real" curved trajectory. This process, as it is well known, gives the definition of velocity as the limit:

$$v = \frac{ds}{dt} = \lim_{\Delta t \to 0} \frac{\Delta s}{\Delta t} = \frac{dr}{dt} \tag{1.4}$$

Notice that r, v and Δs are vectors. The momentum p is then:

$$p = mv \tag{1.5}$$

---

[6]This law is also known as Galileo's principle or law of inertia.

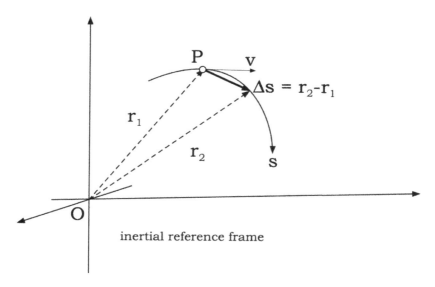

Figure 1.1 Geometry to illustrate the concept of velocity for a body moving along a trajectory s.

eq. 1.3 then becomes:

$$F = \frac{d}{dt}(mv) \tag{1.6}$$

In the case of constant mass m, we have:

$$F = m\frac{dv}{dt} = m\frac{d^2r}{dt^2} \tag{1.7}$$

if we know the mass of a particle and the force acting on it, then we can calculate its motion by solving the second-order differential equation 1.7 above. Eq. 1.6 tells us the first important conservation law: if the force F acting on a particle is zero, then the momentum is conserved, i.e. is constant with time.

We can obtain another important conservation law by vector multiplication of equation 1.6 by r:

$$r \times F = r \times \frac{d}{dt}(mv) \tag{1.8}$$

We now use the property of the derivative of a vector product:

$$\frac{d}{dt}(a \times b) = a \times \frac{db}{dt} + \frac{da}{dt} \times b \tag{1.9}$$

Using eq. 1.9 on eq. 1.8 we have:

$$\frac{d}{dt}(r \times mv) = \frac{dr}{dt} \times mv + r \times \frac{d}{dt}(mv) = v \times mv + r \times \frac{d}{dt}(mv) \tag{1.10}$$

we know that the vector product of a vector with itself is zero, and therefore, using eq. 1.8, we can write:

$$r \times F = \frac{d}{dt}(r \times mv) \tag{1.11}$$

Eq. 1.11 tells us a few interesting things. First, let's identify the various expressions in eq. 1.11. The quantity $r \times F$ is called **moment** of the Force F around the point O origin of the vector r. The moment of a force is also called **torque**. More in general, any physical quantity can be multiplied by a distance to produce a moment.

The right-hand side term of eq. 1.11 contains the time derivative of the momentum mv multiplied by the vector r. This quantity is called **angular momentum** $L = r \times mv$ which can be interpreted and the moment of the momentum $p = mv$. The angular momentum is therefore conserved whenever the left-hand side of eq. 1.11 is zero. This happens when the force $F = 0$ or when the vector product $r \times F = 0$. A particularly interesting case of conservation of angular momentum is held when the force F is a force always directed along the vector r, such as in the case of so-called **central forces**. In fact, $r \times F$ is null when the two vectors r and F are parallel. Newton's law of gravity is an example of such a radial force, and therefore, all orbiting systems bound by gravity conserve angular momentum.

1.2 LAGRANGE

The Newtonian equations of motion of the previous section are mostly relationships between vectors. As such, the equations are valid only if we restrict ourselves to *inertial* frames. It would be quite useful if we could express the equation of motions in a formalism that is valid in any reference frame and not just the inertial ones. Scalar quantities, for example, are by definition, the same in all reference frames. Would it be possible to build a formalism in which we deal just with scalars (or combination of scalars)?

In order to do so, we need to go back to Hero of Alexandria (10 AD to 70 AD) [43]. This Greek philosopher is probably the first to consider the search for a *minimum* as a fundamental principle to interpret natural phenomena. In his book *Catoptrics* he showed that light reflected from a mirror takes the possible shortest path from the object to the observer. We will see that this idea of a fundamental principle from which we can derive the laws of nature is at the basis of modern physics.

Fermat, a French mathematician (1601-1665), wanted to see if it is possible to use Hero's idea of a minimum path to include not just mirror's reflection but also refraction, i.e. transition of the light rays between media with different refraction indexes.

In fig. 1.2 a light ray travels from the point P, immersed in a medium of index of refraction n, to the point Q at the interface with a different medium with index of refraction $n' > n$. From Q the light ray then reaches the point R. We know that in a medium with index of refraction n light travels with a speed $v = c/n$, where c is the speed of light in vacuum. Fermat immediately realized that the light ray does not follow a path of minimum distance $PQ'R$. We can try other ideas like, for example,

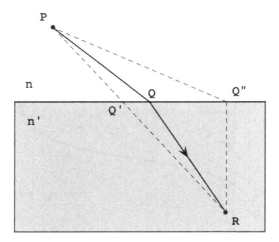

Figure 1.2 Fermat's principle. The light ray propagates from P (in medium of index of refraction n) to R (in medium of index of refraction n') through the path of minimum time PQR instead of minimum length $PQ'R$. $PQ''R$ is another possible path.

the path $PQ''R$, where the light spends a lot of time where it is fast and less time where it is slower. Also in this case, the light ray follows a different path.

Fermat found that light follows the path of "minimum time": the light ray goes from P to R along the path that minimizes the time needed to reach R. It can be shown that the minimum time for a light ray to go from P to Q is:

$$\tau = \frac{1}{c}(\overline{PQ} \cdot n + \overline{QR} \cdot n') \tag{1.12}$$

where c is the speed of light in vacuum.

This is the famous Fermat's principle in its strong form, i.e. requiring a minimum time[7]. A more accurate statement requires the light ray to propagate along a trajectory that is *stationary*, i.e. its first-order variation is zero.

Let's briefly discuss this idea of stationary path for a light ray. Of all the possible paths, light propagates along the stationary path, i.e. the path for which its variation is zero at the first order. If we want to use calculus, we know how to calculate the derivative of a function: we know that imposing such derivative to vanish implies that we are looking for points that are either on a minimum, a maximum or a saddle point. But we are not looking for a special point: we are looking for a special path or, in other words, for a special function. Our problem is more like finding a stationary point of a function of functions (see fig. 1.3). In order to do so, we associate a

---

[7]It can be shown [8] that there are rare special situations in which the light ray actually propagates along a path of *maximum* time.

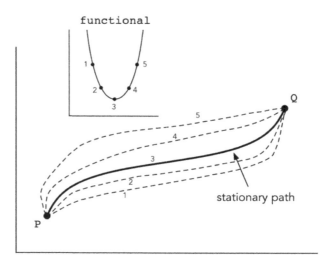

Figure 1.3 Schematic representation of the concept of functional: to each possible path from point P to point Q in space, it is associated with a number. The functional is the function built with all the possible paths. The stationary path (labeled 3) is the one that minimizes the functional.

number to each path[8] and find a stationary point for this function of functions. Function of functions are referred as *functionals*, and the study of its variations is called *variational calculus*.

We can create any functional that we want. An example of functional could be the length of the paths, i.e. a function $f = f(p)$ which returns the length of the path p (a number) if we input a path (a function). Another functional could be the square of the length of the path or its logarithm. In the case of Fermat's principle, and with reference to fig. 1.2, we build the functional:

$$T = \int_P^R dt = \int_P^R \frac{ds}{v} \tag{1.13}$$

The function T is such that if we input a certain path p it returns the time the light needed to traverse it. The points P and Q are two fixed points in space and ds is the element of length along the path p traversed at velocity v. Fermat's principle can be mathematically written as $\delta T = 0$. Where δ indicates the operation of finding the path such that a minimum deviation from it does not change the time to traverse it to the first order. In a slightly more rigorous form, the time required for a light ray to run across a neighboring path differs from the stationary path by a second-order quantity.

---

[8]Which is itself a function.

1.2.1 STATIONARY ACTION AND EULER-LAGRANGE EQUATIONS

Having seen that Fermat's principle describes very well the light propagation, it is natural to ask ourselves whether we can find a similar principle for the motion of particles. In other words, is it possible to obtain Newtonian's mechanics from a stationary principle? Can we find a functional for the motion of matter particles? The answer is yes: there is a functional, called *action*, that returns a number for each possible path followed by a moving particle. The particle will follow the path that makes the action stationary. There are important differences with respect to Fermat's principle: we assume that the particle starts its motion at position 1 at time t_1 and ends its motion at position 2 at time t_2. All the possible potential paths are indicated by the function $x(t)$. We also assume that the action integral contains the function $\mathscr{L} = \mathscr{L}(x(t),\dot{x}(t))$, called *Lagrangian*[9].

The action is:

$$S = \int_{t_1}^{t_2} \mathscr{L}(x(t),\dot{x}(t))dt \tag{1.14}$$

and, in analogy with Fermat's principle, we search for the path for which $\delta S = 0$. This principle is referred to as *principle of least action* or *Hamilton's principle*[10] with the *caveat* that it is more correct to refer to stationary action rather than least action.

For completeness, there is another stationary principle called *Maupertuis's* principle where the following integral is stationary:

$$S = \int_{q_1}^{q_2} p dq \tag{1.15}$$

where p is the momentum and S is the action. Eq. 1.15 was first written by Euler. Eq. 1.15 is very interesting because it refers to *generalized coordinates* q and p, not necessarily position x and momentum mv. The concept of generalized coordinate is used often and in different contexts in the rest of the book. Therefore, it is very important that we define better the concept.

In general, in classical mechanics, if we want to completely describe the evolution of a system composed of N particles in a Cartesian coordinate system, we need to provide a set of $6N$ parameters: $3N$ spatial coordinates x_i and $3N$ velocity coordinates v_i, where $i = 1, 2, ..., N$. In the Lagrangian formalism we are not bound to Cartesian coordinates: any set of coordinates q_i that uniquely specify the configuration of the system under study is allowed.

We know that the equation of motion is second-order differential equations, and therefore, we need to specify the associated momenta p_i. We are left to identify this Lagrangian function and what conditions it has to be subject to correctly describe the

---

[9]From the Italian mathematician and astronomer Joseph-Luis Lagrange, born in Turin with the name Giuseppe Luigi Lagrangia in 1736

[10]From Sir W. R. Hamilton, Irish mathematician (1805-1865) has formulated the principle using the Lagrangian. Pierre Louis Maupertuis is also credited for being one of the first to propose the principle in different formulation together with Euler and Leibniz.

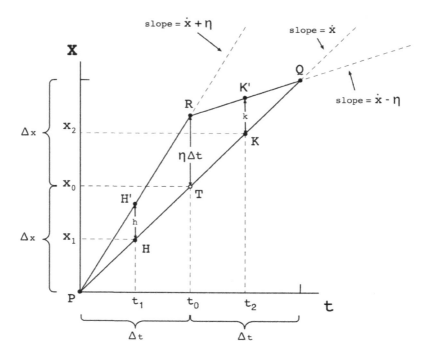

Figure 1.4 Geometric construction for obtaining the Euler-Lagrange equations.

motion. For simplicity, we will now give a geometrical derivation of the so-called *Euler-Lagrange* equations for the motion of a single particle, i.e. the equations that the associated Lagrangian function has to satisfy in order to properly describe the trajectory of such particle. From what we have discussed above, we assume that the trajectory is such that the action in eq. 1.14 is stationary.

With reference to fig. 1.4, let us consider a particle going from point P to point Q through two different trajectories: $\overline{PTQ}$ and a slightly perturbed $\overline{PRQ}$. For simplicity, we assume that the motion happens in just one coordinate x so that $\mathscr{L} = \mathscr{L}(x, \dot{x})$. Fig. 1.4 is depicting the motion of a particle from P to T and from T to Q at constant speed $\dot{x}$ or alternatively from P to R at slightly higher speed $\dot{x} + \eta$ and from R to Q at a slightly slower speed $\dot{x} - \eta$ such that both paths are traversed with the same time $t_1 + t_2 = 2\Delta t$. We consider all the segments in fig. 1.4 as *infinitesimal*.

The reasoning now goes as follows: we calculate the action along the straight-line trajectory S_{PTQ} which we already know is stationary because we know that a particle will follow a straight line trajectory[11]. We calculate then the action $S_{PRQ} = S_{PR} + S_{RQ}$

---

[11]We are assuming that there are no external forces acting upon the particle.

along the infinitesimally close path. By equating $S_{PTQ} = S_{PRQ}$ we obtain a condition for the Lagrangian $\mathscr{L}$ to be stationary.

The action along the straight line $\overline{PTQ}$ is calculated immediately by noticing that the point T divides the segment $\overline{PQ}$ exactly in half by construction. It follows that the action integral 1.14 is:

$$S_{PTQ} = \int_{t_1}^{t_2} \mathscr{L}(x(t),\dot{x}(t))dt \approx 2\Delta t \mathscr{L}(x_0,\dot{x}_0) \qquad (1.16)$$

where $\mathscr{L}(x_0,\dot{x}_0)$ is the average value of the Lagrangian at $x = x_0$.

We now need to calculate the action S_{PRQ} which is equal to the sum of the action S_{PR} plus the action S_{RQ}. In order to calculate the action S_{PR}, we need to calculate the Lagrangian $\mathscr{L}(H')$ at the point H' which, in turn, can be approximated by a Taylor's expansion of the Lagrangian $\mathscr{L}(H)$ around the point H. We have:

$$\mathscr{L}(H') \sim \mathscr{L}(H) + \Delta x \frac{\partial \mathscr{L}}{\partial x}\Big|_{t=t_1} + \Delta \dot{x} \frac{\partial \mathscr{L}}{\partial \dot{x}}\Big|_{t=t_1} + \ldots \qquad (1.17)$$

where Δx is the segment $\overline{HH'}$ and $\Delta \dot{x}$ is the difference between the slope at H, equal to $\dot{x}$, and the slope at H', equal to $\dot{x} + \eta$. It follows that $\Delta \dot{x}$ in eq. 1.17 is η. From fig. 1.4 it is easy to verify that the two triangles $\triangle PRT$ and $\triangle PHH'$ are similar and therefore $\Delta x = \frac{1}{2}\eta\Delta t$. Using eq. 1.16, the action S_{PR} can be approximated by:

$$S_{PR} \approx \Delta t \left[\mathscr{L}(x_1,\dot{x}_1) + \frac{1}{2}\eta\Delta t \frac{\partial \mathscr{L}}{\partial x}\Big|_{t=t_1} + \eta \frac{\partial \mathscr{L}}{\partial \dot{x}}\Big|_{t=t_1} \right] \qquad (1.18)$$

where, in analogy to eq. 1.16, the action is approximated by averaging over the line $\overline{PR}$ for a time interval Δt.

Similarly, the action S_{RQ} is approximated by averaging over the segment $\overline{RQ}$:

$$S_{RQ} \approx \Delta t \left[\mathscr{L}(x_2,\dot{x}_2) + \frac{1}{2}\eta\Delta t \frac{\partial \mathscr{L}}{\partial x}\Big|_{t=t_2} - \eta \frac{\partial \mathscr{L}}{\partial \dot{x}}\Big|_{t=t_2} \right] \qquad (1.19)$$

We can now add the two actions 1.18 and 1.19 to obtain the action of the perturbed trajectory $\overline{PRQ}$:

$$\begin{aligned} S_{PR} + S_{RQ} \approx &\ \Delta t \left[\mathscr{L}(x_1,\dot{x}_1) + \mathscr{L}(x_2,\dot{x}_2) \right] \\ &+ \frac{1}{2}\eta\Delta t^2 \left[\frac{\partial \mathscr{L}}{\partial x}\Big|_{t=t_1} + \frac{\partial \mathscr{L}}{\partial x}\Big|_{t=t_2} \right] \\ &+ \eta\Delta t \left[\frac{\partial \mathscr{L}}{\partial \dot{x}}\Big|_{t=t_1} - \frac{\partial \mathscr{L}}{\partial \dot{x}}\Big|_{t=t_2} \right] \end{aligned} \qquad (1.20)$$

The sum of the first two terms of eq. 1.20 can be expressed as averages in virtue of eq. 1.16 and they are therefore equal to the average along the whole segment $\overline{PQ}$ being the sum of the two averages at H and K. A similar argument is valid for the other terms.

The last term in eq. 1.20 is the difference in the function between two points that are close in time, which is approximated by the derivative with respect to the time of the function, times the infinitesimal time interval Δt.

We can therefore write:

$$
\begin{aligned}
S_{PR} + S_{RQ} &\approx \Delta t \left[2\mathscr{L}(x_0, \dot{x}) + \eta \Delta t \frac{\partial \mathscr{L}}{\partial x} \Big|_{t=t_0} - \eta \Delta t \frac{d}{dt} \frac{\partial \mathscr{L}}{\partial \dot{x}} \Big|_{t=t_0} \right] \\
&= \Delta t \left[2\mathscr{L}(x_0, \dot{x}) + \eta \Delta t \left(\frac{\partial \mathscr{L}}{\partial x} \Big|_{t=t_0} - \frac{d}{dt} \frac{\partial \mathscr{L}}{\partial \dot{x}} \Big|_{t=t_0} \right) \right]
\end{aligned}
\tag{1.21}
$$

Comparing eq. 1.16 with eq. 1.21 we see that the two actions, in order to be equal, must satisfy the following equation:

$$
\frac{\partial \mathscr{L}}{\partial x} \Big|_{t=t_0} - \frac{d}{dt} \frac{\partial \mathscr{L}}{\partial \dot{x}} \Big|_{t=t_0} = 0
\tag{1.22}
$$

Eq. 1.22 is known as the **Euler-Lagrange** equation.

More in general, the Euler-Lagrange is a set of equations referred to *generic* coordinates q and their time derivatives $\dot{q}$. Therefore, a more general Euler-Lagrange equation can be written as:

$$
\frac{\partial \mathscr{L}}{\partial q_i} - \frac{d}{dt} \frac{\partial \mathscr{L}}{\partial \dot{q}_i} = 0
\tag{1.23}
$$

where the index $i = 1, .., N$ runs over the N independent degrees of freedom (DOF), i.e. the number of independent parameters needed to uniquely identify the state (or configuration) of the mechanical system[12].

We are left to find this special function $\mathscr{L}$, the Lagrangian, such that we can recover the motion of the particle(s). It turns out that the correct choice to recover correctly the motion of particles is:

$$
\mathscr{L} = T - V
\tag{1.24}
$$

where T is the kinetic energy and V is the potential energy. There is not a specific reason why this particular combination of kinetic and potential energy gives the proper Lagrangian.

It is instructive to show that the Euler-Lagrange equations keep their form in any coordinate system. First, we show that in Cartesian coordinates, Euler-Lagrange equations are equal to Newton's $F = ma$. For simplicity let's consider the motion of a particle in a Cartesian coordinate system x where the motion is constrained on a line. In this coordinate system the Lagrangian is:

$$
\mathscr{L} = T - V = \frac{1}{2} m \dot{x}^2 - V(x)
\tag{1.25}
$$

---

[12] For example, a rigid body in space, not subject to any constraint, has 6 degrees of freedom: 3 because it can translate along three independent directions and 3 because it can rotate around three independent axis of rotation.

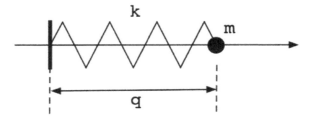

Figure 1.5 Simple harmonic oscillator.

where $V(x)$ is a scalar function representing the potential energy depending only on the coordinate x. Let's calculate the two terms of Euler-Lagrange equation (in this case it is just one equation):

$$\frac{\partial \mathcal{L}}{\partial x} = \frac{dV}{dx}$$
$$\frac{d}{dt}\frac{\partial \mathcal{L}}{\partial \dot{x}} = m\ddot{x}$$

(1.26)

It follows that eq. 1.23 for just the coordinate x becomes:

$$m\ddot{x} = -\frac{dV}{dx}$$

(1.27)

If the force F can be expressed as the derivative of a scalar potential function, then eq. 1.27 is exactly Newton's equation $F = ma$.

The simple classical harmonic oscillator is another useful example of Lagrangian formalism that we will need later in the book. Let us assume we have a point mass m forced to move on a straight line of coordinate q and connected to a spring of elastic constant k working in linear regime, i.e. the force experienced by the point mass, in compression or extension, is $F = -kq$, i.e. linear in q (see fig. 1.5). It is well known that the linearity is a good approximation for small q when deformations of the spring are perfectly elastic. For this simple physical system, the kinetic energy is $T = 1/2m\dot{q}^2$ while the potential energy is $V = 1/2kq^2$. The Lagrangian is:

$$\mathcal{L} = T - V = \frac{1}{2}m\dot{q}^2 - \frac{1}{2}kq^2 = \frac{1}{2}m\dot{q}^2 - \frac{1}{2}m\omega^2 q^2$$

(1.28)

where $\omega = 2\pi\nu = \sqrt{k/m}$ is the angular frequency of the oscillations of the mass m. Applying the Euler-Lagrange equations to eq.1.28 we obtain the differential equation for the motion of the mass m:

$$\frac{d}{dt}\frac{\partial \mathcal{L}}{\partial \dot{q}} - \frac{\partial \mathcal{L}}{\partial q} = \ddot{q} + \omega^2 q = 0$$

(1.29)

which has the solution:

$$q = A\sin(\omega t + \phi)$$

(1.30)

showing clearly the sinusoidal motion of the mass m around its equilibrium position.

We now ask the question: to describe a simple harmonic oscillator how many degrees of freedom do we need? Let's first refresh what the degrees of freedom (DOF) is: in a mechanical system, the DOFs are the number of parameters needed to define its state. A rigid body in a 3-dimensional space, for example, has 6 DOFs: 3 parameters are needed to define the position of its center of mass and 3 parameters are needed to define the 3 rotation angles. We will see in the next section that the Lagrangian formalism is not limited to a particular choice of coordinates and therefore, no matter what coordinate system we choose, we still have 6 DOFs. A simple harmonic oscillator needs two parameters to be completely described: they can be its position and momentum or its kinetic and potential energy and so it has 2 degrees of freedom. After all, we just learned that we can obtain the equation of motion by writing the Lagrangian which is exactly the difference between the kinetic and the potential energy thus showing clearly the interchangeability of position/momentum and kinetic/potential energy as a couple of parameters.

1.2.2 CHANGE OF COORDINATES

Lagrangian formalism has two main advantages with respect to the original Newtonian formalism of $F = ma$: it uses *scalar quantities* instead of vectors and has equations that keep their form in any coordinate system. Euler-Lagrange equation have the great advantage to express the dynamics without resorting to vectors which is undoubtedly an advantage when the systems are complicated. Another advantage comes from studying Euler-Lagrange equation in other coordinate systems instead of Cartesian. Let's prove that Euler-Lagrange equations keep their form in *any* coordinate system.

Let us suppose that $(x_i = x_1, x_2, ..., x_N)$ is a coordinate system – for example, Cartesian – for which eq. 1.23 is written:

$$\frac{\partial \mathscr{L}}{\partial x_i} - \frac{d}{dt}\frac{\partial \mathscr{L}}{\partial \dot{x}_i} = 0 \qquad (1.31)$$

and suppose that we have a new coordinate system:

$$q_i = q_i(x_1, x_2, ..., x_N) = q_i(x_i) \qquad (1.32)$$

and we assume that we can invert eq. 1.32:

$$x_i = x_i(q_1, q_2, ..., q_N) = x_i(q_i). \qquad (1.33)$$

We want to show that if eq. 1.31 holds, then the same equation holds when we substitute x_i and $\dot{x}_i$ with q_i and $\dot{q}_i$. In order to do so, let's study the following partial derivatives of $\mathscr{L}(q_i, \dot{q}_i)$:

$$\frac{\partial \mathscr{L}}{\partial \dot{q}_j} = \sum_{i=1}^{N} \frac{\partial \mathscr{L}}{\partial \dot{x}_i} \frac{\partial \dot{x}_i}{\partial \dot{q}_j} \qquad (1.34)$$

which is obtained by using the chain rule on each of the j components, with $j = 1, 2, ..., N$. Using the chain rule on eq. 1.33 we have:

$$\dot{x}_i = \sum_{i=1}^{N} \frac{\partial x_i}{\partial \dot{q}_j} \dot{q}_j \tag{1.35}$$

a straight differentiation of eq. 1.35 gives:

$$\frac{\partial \dot{x}_i}{\partial \dot{q}_j} = \frac{\partial x_i}{\partial q_j} \tag{1.36}$$

and eq. 1.34 then becomes:

$$\frac{\partial \mathscr{L}}{\partial \dot{q}_j} = \sum_{i=1}^{N} \frac{\partial \mathscr{L}}{\partial \dot{x}_i} \frac{\partial x_i}{\partial q_j}. \tag{1.37}$$

The time derivative of eq.1.37 gives:

$$\begin{aligned}
\frac{d}{dt} \left[\frac{\partial \mathscr{L}}{\partial \dot{q}_j} \right] &= \frac{d}{dt} \left[\sum_{i=1}^{N} \frac{\partial \mathscr{L}}{\partial \dot{x}_i} \frac{\partial x_i}{\partial q_j} \right] \\
&= \sum_{i=1}^{N} \frac{d}{dt} \left(\frac{\partial \mathscr{L}}{\partial \dot{x}_i} \right) \frac{\partial x_i}{\partial q_j} + \sum_{i=1}^{N} \frac{\partial \mathscr{L}}{\partial \dot{x}_i} \frac{d}{dt} \frac{\partial x_i}{\partial q_j}.
\end{aligned} \tag{1.38}$$

We now switch the order of differentiation of $\frac{d}{dt}$ with $\frac{\partial}{\partial q_j}$ in the second term of eq. 1.38[13]:

$$\begin{aligned}
\frac{d}{dt} \left[\frac{\partial \mathscr{L}}{\partial \dot{q}_j} \right] &= \sum_{i=1}^{N} \frac{d}{dt} \left(\frac{\partial \mathscr{L}}{\partial \dot{x}_i} \right) \frac{\partial x_i}{\partial q_j} + \sum_{i=1}^{N} \frac{\partial \mathscr{L}}{\partial \dot{x}_i} \frac{d}{dt} \frac{\partial x_i}{\partial q_j} \\
&= \sum_{i=1}^{N} \frac{d}{dt} \left(\frac{\partial \mathscr{L}}{\partial \dot{x}_i} \right) \frac{\partial x_i}{\partial q_j} + \sum_{i=1}^{N} \frac{\partial \mathscr{L}}{\partial \dot{x}_i} \frac{\partial \dot{x}_i}{\partial q_j}.
\end{aligned} \tag{1.39}$$

Using eq. 1.31, eq. 1.39 can be written as:

$$\begin{aligned}
\frac{d}{dt} \left[\frac{\partial \mathscr{L}}{\partial \dot{q}_j} \right] &= \sum_{i=1}^{N} \left(\frac{\partial \mathscr{L}}{\partial x_i} \right) \frac{\partial x_i}{\partial q_j} + \sum_{i=1}^{N} \frac{\partial \mathscr{L}}{\partial \dot{x}_i} \frac{\partial \dot{x}_i}{\partial q_j} \\
&= \frac{\partial \mathscr{L}}{\partial q_j}
\end{aligned} \tag{1.40}$$

which are exactly the Euler-Lagrange equations for the new coordinates q_j.

---

[13]This is the so-called *symmetry of the second derivative* or, sometimes, Clairaut'as theorem or Schwarz's theorem. This theorem holds for any point P around which the second partial derivatives are continuous.

We have seen that Newton's equation are *vector* equations valid only in inertial reference frames. We did not show it, but we used a restricted form of coordinate transformation in eq. 1.32.

More in general, Euler-Lagrange equations are valid even if there is a time dependence in the coordinate transformation of the form:

$$q_i = q_i(x_1, x_2, ..., x_N, t) = q_i(x_i, t) \tag{1.41}$$

This generalization is useful, for example, when we consider the physics of a rotating Cartesian coordinate system (x', y', z'), rotating around the z axis with angular velocity ω, with respect to a fixed Cartesian coordinate system (x, y, z).

Let's consider a free particle whose Lagrangian is:

$$\mathscr{L} = \frac{1}{2} m \dot{r}^2 \tag{1.42}$$

where $r = x^2 + y^2 + z^2$ is the cartesian coordinate. Applying the Euler-Lagrange equations to this Lagrangian will obviously get us $m\ddot{r} = 0$. We can integrate once to give $m\dot{r} = const$, meaning that the particle of mass m is traveling at constant speed.

Let's now change coordinate system, from the fixed (x, y, z) to the rotating (x', y', z'). We know that the form of Euler-Lagrange will not change in the new coordinates. However, the Lagrangian will change and so we expect to find different equation of motions. We should not be surprised because the motion of a free particle will certainly look different if we are rotating together with the rotating coordinate system. Let's first write down the coordinate transformation:

$$\begin{aligned} x' &= x \cos \omega t - y \sin \omega t \\ y' &= x \sin \omega t + y \cos \omega t \\ z' &= z \end{aligned} \tag{1.43}$$

After a little bit of algebra, the Lagrangian 1.42 becomes, in the new coordinate system:

$$\mathscr{L}' = \frac{1}{2} m \left[\dot{r}' + \omega \times r' \right]^2 \tag{1.44}$$

and the Euler-Lagrange equations with respect to $\dot{r}'$ and r' give:

$$m(\ddot{r}' + \omega \times (\omega \times r') + 2\omega \times \dot{r}') = 0 \tag{1.45}$$

where the second and third term are, respectively, the well known centrifugal and Coriolis fictitious forces.

1.2.3 NOETHER THEOREM

In this section we will briefly discuss an important theorem due to Noether[14]. We start by consider three hypothetical experiments where we observe a system described by a Lagrangian $\mathscr{L}(x, \dot{x}, t)$. Note that we are allowing our Lagrangian to have

---

[14]Amalie Emmy Noether (23 March 1882 - 14 April 1935). Considered one of the most important women in the history of mathematics for her work in mathematical physics and abstract algebra.

an explicit dependence from the time t in addition to the position x and its derivative $\dot{x}$.

Let's observe a system at position x and require that if we move the system at a different position $x' = x + \varepsilon$, where ε is an arbitrary constant, the Lagrangian is unaltered. This means that the system has exactly the same equation of motion coming from the Euler-Lagrange equations[15]. There is a very special condition for which this condition is clearly satisfied, i.e. when the Lagrangian does not explicitly depend on the coordinate x : $\mathscr{L}(x,\dot{x},t) = \mathscr{L}(\dot{x},t)$. In this case, we can operate any transformation on the coordinate x and the Lagrangian obviously stays the same not having the explicit dependence on x. When a coordinate does not explicitly appear in the Lagrangian, this coordinate is called *cyclic coordinate*.

If a Lagrangian contains a cyclic coordinate, then the corresponding derivative in Euler-Lagrange equations is zero and Euler-Lagrange equations simplify to:

$$\frac{\partial \mathscr{L}}{\partial x} - \frac{d}{dt}\frac{\partial \mathscr{L}}{\partial \dot{x}} = 0$$
$$\frac{d}{dt}\frac{\partial \mathscr{L}}{\partial \dot{x}} = 0 \qquad\qquad (1.46)$$
$$\frac{\partial \mathscr{L}}{\partial \dot{x}} = \text{const.}$$

We have alreay encountered a Lagrangian with no x dependence: the free particle with $\mathscr{L} = \frac{1}{2}m\dot{x}^2$. Last equation in eq. 1.46 gives:

$$p = m\dot{x} = \frac{\partial \mathscr{L}}{\partial \dot{x}} = \text{const.} \qquad\qquad (1.47)$$

Eq. 1.47 tells us that the momentum p associated with the coordinate x is constant if the Lagrangian is cyclic with respect to x. We have reached the interesting conclusion that if the Lagrangian does not depend on a coordinate x, then its associated momentum p is constant. It seems therefore that the origin of the conservation of momentum of Newtonian mechanics is related to a specific symmetry of the Lagrangian. By "symmetry" here we intend the fact that the Lagrangian is unchanged with respect to an operation which, in this case, is any transformation $x \to x + \varepsilon$ where $\varepsilon =$const. We can extend this idea to the generalized coordinates q_i: if the Lagrangian does not depend explicitly on the generalized coordinate q_i then its associated momentum $p_i = \frac{d\mathscr{L}}{d\dot{q}_i} =$constant.

Let's now study another symmetry of the Lagrangian. Let's require that our Lagrangian is symmetric under any rotation of the coordinate system, i.e. we want that our dynamical equations stay the same if the coordinate system is rotated by a fixed angle θ. Let's limit ourselves to the motion on a 2-dimensional plane x, y. The Lagrangian in Cartesian coordinates is $\mathscr{L} = \mathscr{L}(x,y,\dot{x},\dot{y})$. Rotations of coordinate

---

[15]This is equivalent to simply require that if we perform an experiment in Rome, for example, then the same experiment must give the same results if performed in New York.

systems are better treated if we express the Lagrangian in polar coordinates (r, θ). The Cartesian to Polar coordinate transformations together with its derivatives are expressed by:

$$\begin{aligned}
x &= r\cos\theta \\
y &= r\sin\theta \\
\dot{x} &= \dot{r}\cos\theta - r\dot{\theta}\sin\theta \\
\dot{y} &= \dot{r}\sin\theta + r\dot{\theta}\cos\theta
\end{aligned} \qquad (1.48)$$

the Lagrangian in Cartesian coordinate system is:

$$\mathscr{L} = \frac{1}{2}m(\dot{x}^2 + \dot{y}^2) - V(x,y) \qquad (1.49)$$

where $V(x,y)$ is a scalar function of the coordinates x,y representing the potential energy. Let's restrict to the case where the forces are conservative and the potential function is $V = V(r)$ only a function of r.

Using the coordinate transformations 1.48, eq. 1.49 becomes:

$$\begin{aligned}
\mathscr{L} &= \frac{1}{2}m(\dot{x}^2 + \dot{y}^2) - V(r) \\
&= \frac{1}{2}m(\dot{r}^2 + r^2\dot{\theta}^2) - V(r)
\end{aligned} \qquad (1.50)$$

Eq. 1.50 is cleary cyclic with respect to the coordinate θ which means that the associated momentum is conserved. The associated momentum p_θ is:

$$\begin{aligned}
p_\theta &= \frac{\partial}{\partial\dot{\theta}}\left[\frac{1}{2}m(\dot{r}^2 + r^2\dot{\theta}^2) - V(r)\right] \\
&= mr^2\dot{\theta} = \text{constant}
\end{aligned} \qquad (1.51)$$

which represents the *conservation of angular momentum*. A Lagrangian symmetric with respect to transformations $\theta \rightarrow \theta + \theta_0$, where θ_0 is an arbitrary fixed rotation, implies the conservation of angular momentum.

The last symmetry left to discuss is the time shifts. We want to study what are the consequences of requiring that the Lagrangian is invariant with respect to transformations like $t \rightarrow t + t_0$ where t_0 is an arbitrary amount of time. This requirement broadly translates into asserting that if my system is subject to certain dynamical equations today, the same equations must hold yesterday or tomorrow or at any time $t + t_0$.

Time invariance of the Lagrangian $\mathscr{L}(x,\dot{x},t)$ means that its time derivative must be zero. Alternatively, we say that the Lagrangian does not have *explicit* dependence on t so that the Lagrangian is $\mathscr{L}(x,\dot{x})$.

Using the chain rule:

$$\begin{aligned}
\frac{d}{dt}\mathscr{L}(x,\dot{x}) &= \frac{d\mathscr{L}}{dx}\dot{x} + \frac{d\mathscr{L}}{d\dot{x}}\ddot{x} \\
&= \left[\frac{d}{dt}\frac{\partial\mathscr{L}}{d\dot{x}}\right] + \frac{d\mathscr{L}}{d\dot{x}}\ddot{x}
\end{aligned} \qquad (1.52)$$

where we have used the Euler-Lagrange equation. Eq. 1.52 implies:

$$\frac{d}{dt}\left[\dot{x}\frac{\partial \mathscr{L}}{\partial \dot{x}} - \mathscr{L}\right] = 0 \tag{1.53}$$

the quantity in parenthesis is conserved if the Lagrangian is time independent. It is easy to show that this conserved quantity is the total energy $T + V$. The expression in parenthesis in eq. 1.53, with a Lagrangian $\mathscr{L} = T - V = 1/2m\dot{x}^2 - V(x)$:

$$\begin{aligned} E &= \dot{x}\frac{\partial \mathscr{L}}{\partial \dot{x}} - \mathscr{L} \\ &= m\dot{x}^2 - (T - V) \\ &= 2T - (T - V) \\ &= T + V \end{aligned} \tag{1.54}$$

which is the well-known mechanical energy expressed as the sum of the kinetic and potential energy. Therefore, conservation of energy follows from the invariance of the Lagrangian under time translation.

Returning to the generalized coordinates, we define this quantity E as *canonical energy*:

$$E(q,\dot{q}) = \dot{q}\frac{\partial \mathscr{L}}{\partial \dot{q}} - \mathscr{L} \tag{1.55}$$

The canonical energy E is conserved when the corresponding Lagrangian does not depend on time. It should be pointed out that it is possible that the canonical energy is conserved, but it is not the mechanical energy.

The three examples above are three loose applications of a general theorem by Noether which loosely states: "To each continuous symmetry of the Lagrangian there is a corresponding quantity which is time independent"[16]. A full demonstration of Noether's theorem is beyond the scope of this book, and we refer the interested reader to many excellent textbooks like Arnold [2].

1.3 TWO-BODY PROBLEM

As an application of the Lagrangian formalism let us discuss a very famous problem: the dynamics of a small mass m orbiting around a large mass M where $m \ll M$. Let's write the gravitational potential:

$$V(r) = -\frac{GmM}{r} \tag{1.56}$$

---

[16]A more rigorous enunciation given by Arnold [2] would be: "to every one-parameter group of diffeomorphisms of the configuration manifold of a lagrangian system which preserves the lagrangian function, there corresponds a first integral of the equations of motion."

where $G = 6.674 \cdot 10^{-11}$ m^3 kg^{-1} s^{-2} is Newton's constant. The Lagrangian for the two body problem, in polar coordinates, is obtained by using the potential 1.56 into eq. 1.50:

$$\mathscr{L} = \frac{1}{2}m(\dot{r}^2 + r^2\dot{\theta}^2) - V(r) = \frac{1}{2}m(\dot{r}^2 + r^2\dot{\theta}^2) + \frac{GmM}{r} \tag{1.57}$$

We now apply the Euler-Lagrange equations 1.23 for the two polar coordinates θ and r:

$$\frac{\partial \mathscr{L}}{\partial \theta} - \frac{d}{dt}\frac{\partial \mathscr{L}}{\partial \dot{\theta}} = 0$$
$$\frac{\partial \mathscr{L}}{\partial r} - \frac{d}{dt}\frac{\partial \mathscr{L}}{\partial \dot{r}} = 0 \tag{1.58}$$

The first equation in 1.58 tells us that momentum is conserved. In fact, the Lagrangian 1.57 is cyclic with respect to the θ coordinate. By Noether theorem the associated angular momentum L is conserved:

$$L = \frac{\partial \mathscr{L}}{\partial \dot{\theta}} = mr^2\dot{\theta} = \text{const.} \tag{1.59}$$

The second equation in 1.58 gives:

$$m\ddot{r} - mr\dot{\theta}^2 = -\frac{GmM}{r^2} \tag{1.60}$$

Using eq. 1.59 and indicating $\ell = \frac{L}{m}$ the angular momentum per unit mass, we eliminate θ to have a differential equation for the coordinate r:

$$m\left(\ddot{r} - \frac{\ell^2}{r^3}\right) = -\frac{GmM}{r^2} \tag{1.61}$$

This differential equation is rather difficult to solve in this form. However, it can be shown that, under a suitable change of functions, the equation can be simplified to give as a solution:

$$r = \frac{\frac{\ell^2}{GM}}{1 + e\cos\theta} \tag{1.62}$$

where e, the eccentricity of the orbit, represents how much the orbit deviates from a circle ($e = 0$ for circular orbit; $0 < e < 1$ for elliptical orbit; $e = 1$ for parabolic orbit and $e > 1$ for hyperbolic orbit). The mathematical details are reported in all advanced books of classical mechanics (see, for example, [35]).

The total conserved mechanical energy of an object of mass m orbiting an object of much larger mass M is:

$$E = \frac{1}{2}m\dot{r}^2 + \frac{1}{2}\frac{L^2}{mr^2} - \frac{GmM}{r} \tag{1.63}$$

Eq. 1.63 shows that the total energy is the sum of three components which are, respectively: kinetic energy, a term containing the angular momentum and the potential energy. The gravitational potential energy can be expressed as the derivative of a

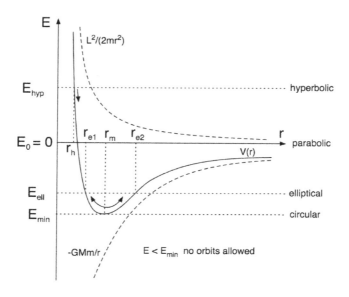

Figure 1.6 Relationship between orbits and total mechanical energy.

conserved force. In complete analogy, the second term in eq. 1.63 can be expressed as a potential U_c of the form:

$$U_c = \frac{1}{2} \frac{L^2}{mr^2} \tag{1.64}$$

whose derivative produces an associated force of the form:

$$F_c = -\frac{\partial U_c}{\partial r} = \frac{L^2}{mr^3} = mr\dot{\theta}^2 \tag{1.65}$$

The force 1.65 is called *centrifugal force* as it is directed from the large mass M to the small mass m exactly opposite to the gravitational attraction on the small mass m.

We can regard the potential energy 1.64 as centrifugal potential energy. If we include it with the gravitational potential energy 1.56 we have an *effective potential energy* of the form:

$$V(r) = -\frac{GmM}{r} + \frac{L^2}{2mr^2} \tag{1.66}$$

The effective potential $V(r)$ of eq. 1.66 is plotted as a solid line in fig. 1.6. Its two components are plotted as dashed lines. The object with mass m will be orbiting the more massive object of mass M if its total energy is $E < 0$, i.e. negative. The orbit associated with the minimum allowed energy is the circular orbit with radius r_m. Higher negative energies are associated with more generic elliptical orbits between the two radii r_{e1} and r_{e2}. Any energy $E > E_{min}$ is allowed. When the energy is $E \geq 0$ then the object is not bound anymore and it orbits on a parabolic ($E = 0$) or hyperbolic ($E > 0$) trajectory.

1.4 HAMILTON

Let us now study the form of definition 1.55. We can rewrite eq. 1.55 by noticing that we can introduce the generalized momentum $p = \frac{\partial \mathscr{L}}{\partial \dot{q}}$ (see eqs. 1.47 and 1.51). Let's introduce a new function, the Hamiltonian H, defined as:

$$H(q,p) = \dot{q}p - \mathscr{L} \tag{1.67}$$

Eq. 1.67 is an example of a *Legendre Transform*, i.e. a special kind of transformation that allows to transform a function of certain variables into another function of *conjugate* variables. For simplicity, let us restrict to functions of two variables $f = f(x,y)$ and we want to transform f into a new function $g = g(u,v)$. In our case, we want to find how to transform the Lagrangian $\mathscr{L}(q,\dot{q})$ into $H(q,p)$ where $\dot{q}, p$ are the conjugate variables. It is important to point out that the two functions, if representing physical quantities, will have the same units. In the case of the Lagrangian $\mathscr{L}$, being the difference between kinetic and potential energy, its units will be energy. Therefore H will have units of energy.

Let's write down the differential of the function $f = f(x,y)$:

$$df = \left(\frac{\partial f}{\partial x}\right)_y dx + \left(\frac{\partial f}{\partial y}\right)_x dy \tag{1.68}$$

where the subscripts x, y in eq. 1.68 mean that the partial derivatives are taken keeping, respectively, x and y constant. We can define two new functions u, v:

$$u \equiv \left(\frac{\partial f}{\partial x}\right)_y, \quad v \equiv \left(\frac{\partial f}{\partial y}\right)_x \tag{1.69}$$

where the symbol $\equiv$ means "equal to by definition"[17]. We can rewrite eq. 1.68 as:

$$df = udx + vdy \tag{1.70}$$

Eq. 1.70 shows that (u,x) and (v,y) are the pair of conjugate variables. Let's now write down the differential of the product of the two functions v, y:

$$d(vy) = ydv + vdy \tag{1.71}$$

and compute the following differential obtained by subtracting eq. 1.71 from eq. 1.70:

$$dg = df - d(vy) = udx - vdy \tag{1.72}$$

The function $g = f - vy$ is the Legendre transform of the function f and allows us to change independent variables by building a new function according to eq. 1.72.

---

[17] Other notations include "$\triangleq$" and $\doteq$.

After discussing the Legendre transforms, let's go back to the Hamiltonian. The general Hamiltonian can be written as:

$$H(q_i, p_i) = \sum_{i=1}^{N} \dot{q}_i p_i - \mathscr{L} \tag{1.73}$$

We now derive the so-called *Hamilton's equations*. For simplicity we write the generalized coordinates $(q_i, \dot{q}_i, p_i, \dot{p}_i)$ as $(q, \dot{q}, p, \dot{p})$ omitting the index $i = 1, ..., N$ and dropping the summation over the index i. This implicitly assumes that every time there is the product of two coordinates, they are summed over the index i[18].

Eq. 1.73 can be written in simplified form[19] as:

$$H(q, p) = \dot{q}p - \mathscr{L} \tag{1.74}$$

We start with the generic Lagrangian $\mathscr{L}(q, \dot{q}, t)$ and Hamiltonian $H(q, p, t)$ and we study the differential:

$$d\mathscr{L} = \frac{\partial \mathscr{L}}{\partial q} dq + \frac{\partial \mathscr{L}}{\partial \dot{q}} d\dot{q} + \frac{\partial \mathscr{L}}{\partial t} dt \tag{1.75}$$

using the definition of generalized momentum $p = \frac{\partial \mathscr{L}}{\partial \dot{q}}$, we have:

$$
\begin{aligned}
d\mathscr{L} &= \frac{\partial \mathscr{L}}{\partial q} dq + p d\dot{q} + \frac{\partial \mathscr{L}}{\partial t} dt \\
&= \frac{\partial \mathscr{L}}{\partial q} dq - \dot{q} dp + d(p\dot{q}) + \frac{\partial \mathscr{L}}{\partial t} dt
\end{aligned}
\tag{1.76}
$$

where we used the identity $d(p\dot{q}) = p d\dot{q} + \dot{q} dp$. We can re-arrange the terms in eq. 1.76 and use 1.74:

$$dH = -d(\mathscr{L} - p\dot{q}) = -\frac{\partial \mathscr{L}}{\partial q} dq + \dot{q} dp - \frac{\partial \mathscr{L}}{\partial t} dt \tag{1.77}$$

We can also express the differential of the Hamiltonian $H = H(q, p)$ as:

$$dH = \frac{\partial H}{\partial q} dq + \frac{\partial H}{\partial p} dp + \frac{\partial H}{\partial t} dt \tag{1.78}$$

Eqs. 1.77 and 1.78 are consistent if:

$$\frac{\partial H}{\partial q} = -\frac{\partial \mathscr{L}}{\partial q}, \quad \frac{\partial H}{\partial p} = \dot{q}, \quad \frac{\partial H}{\partial t} = -\frac{\partial \mathscr{L}}{\partial t} \tag{1.79}$$

---

[18] This is equivalent to the famous Einstein's convention over repeated indices.

[19] Eq. 1.67 was written in the same way but it was referring to just one coordinate pair (q,p). In eq. 1.74 we are omitting the index i and the summation over i.

Euler-Lagrange equations $\frac{d}{dt}\frac{\partial\mathscr{L}}{\partial\dot{q}} - \frac{\partial\mathscr{L}}{\partial q}$ can be re-arranged as $\dot{p} = \frac{\partial\mathscr{L}}{\partial q}$. With such substitution, equations 1.79 become:

$$\dot{p} = -\frac{\partial H}{\partial q}, \quad \dot{q} = \frac{\partial H}{\partial p}, \quad \frac{\partial H}{\partial t} = -\frac{\partial\mathscr{L}}{\partial t} \tag{1.80}$$

Eqs. 1.80 are known as *Hamilton's equations*.

Hamilton's equations are $2N$ (coupled) first-order partial differential equations while Euler-Lagrange equations are N second-order partial differential equations.

As an example of Hamiltonian, let's return to the simple harmonic oscillator. Applying a Legendre transform the Lagrangian 1.28, we have:

$$H = p\dot{q} - \mathscr{L} = m\dot{q}^2 - \frac{1}{2}m\dot{q}^2 + \frac{1}{2}m\omega^2 q^2 = \frac{1}{2m}\left(p^2 + m^2\omega^2 q^2\right) \tag{1.81}$$

where we clearly see that the Hamiltonian contains two quadratic terms in p and q corresponding to the two DOFs.

1.5 POISSON BRACKETS

Let's summarize what we have learned so far: we started from Newtonian mechanics and its postulates, based on vectors with dependence of coordinate system. Then Lagrangian where the important quantities are scalar and coordinate independent. Then Hamilton where we can make even more general coordinate transformations where we can mix up q and p. These more general transformations are called *canonical transformations*.

Hamilton's equations tell us that the coordinates q and conjugate momenta p are completely symmetric as evidenced by inspecting eq. 1.80. Describing a system in the Hamiltonian formalism means that we identify the Hamiltonian $H(q,p,t)$, obeying eq. 1.80, where the index i is not explicitly written but refers to the i-th particle composing the system.

We will now study another formalism to describe the time evolution of dynamical systems: the Poisson brackets. In order to do so, let us study an arbitrary function of the generalized coordinates $f = f(q,p,t)$ and let's write its total time derivative:

$$\frac{df}{dt} = \sum_{i=1}^{N}\left(\frac{\partial f}{\partial q_i}\dot{q}_i + \frac{\partial f}{\partial p_i}\dot{p}_i\right) + \frac{\partial f}{\partial t} \tag{1.82}$$

using Hamilton's equations, eq. 1.82 becomes:

$$\frac{df}{dt} = \sum_{i=1}^{N}\left(\frac{\partial f}{\partial q_i}\frac{\partial H}{\partial p_i} - \frac{\partial f}{\partial p_i}\frac{\partial H}{\partial q_i}\right) + \frac{\partial f}{\partial t} \tag{1.83}$$

eq. 1.83 suggests the introduction of a new operation, called Poisson brackets, between two functions $f = f(q,p,t)$ and $g = g(q,p,t)$:

$$\{f,g\} \equiv \sum_{i=1}^{N}\left(\frac{\partial f}{\partial q_i}\frac{\partial g}{\partial p_i} - \frac{\partial f}{\partial p_i}\frac{\partial g}{\partial q_i}\right) \tag{1.84}$$

with such a definition, the time evolution equation 1.83 can be written in a more compact form as:

$$\frac{df}{dt} = \{f,H\} + \frac{\partial f}{\partial t} \tag{1.85}$$

It is easy to show that the Poisson brackets of the canonical coordinates are:

$$\begin{aligned} \{q_i, q_j\} &= 0 \\ \{p_i, p_j\} &= 0 \\ \{q_i, p_j\} &= \delta_{ij}. \end{aligned} \tag{1.86}$$

Poisson brackets satisfy the requirements of *Lie*[20] *algebra*, i.e. an algebra involving a "multiplication rule" that is anticommutative, bilinear and satisfies the Jacobi identity. It can be shown that:

$$\{f,g\} = -\{g,f\}$$
$$\{\alpha\{f_1,g\} + \beta\{f_2,g\}\} = \alpha\{f_1,g\} + \beta\{f_2,g\}\} \tag{1.87}$$
$$\{\{f,g\},h\} + \{\{g,f\},h\} + \{\{h,f\},g\} = 0$$

where the first equation in 1.87 is the anticommutativity, the second is the bilinearity and the third is the Jacobi identity, for any f, g and h. It can be easily shown, and it is left to the reader, to verify that vectors in tridimensional space together with the vector multiplication $u \times v$ is a Lie algebra.

1.6   PHASE SPACE

We conclude this chapter with a brief discussion of the concept of phase space.

We have seen that a dynamical system composed of N particles in 3-dimensional space is completely characterized when we give the 3N coordinates and 3N momenta. More in general, we completely specify the status of a system if we give the p_i generalized momenta and the q_i generalized coordinates.

If we construct an abstract space whose coordinate axis are the p_i and q_i, then the status of a system is represented by a single point. The time evolution, in such a space, is represented by a curve or trajectory as shown in fig. 1.7.

The time evolution is governed by the Hamilton's equations 1.80 discussed in section 1.4. In fact, given a specific initial state, i.e. an initial point in the phase space, Hamilton's equations predict the time evolution of the system. The curve in the phase space is, therefore, the solutions to Hamilton's equations giving the coordinates $q_i = q_i(t)$ and $p_i(t)$. One of the nice features of phase space is that the curves behave in analogy to the flow of a fluid and some related mathematics can be used. For example, the flow curves in the phase space are always incompressible (Liouville's theorem). Another feature is that the Hamiltonian, along these curves, is constant and conserved if the system described is conservative.

---

[20] Sophus Lie (1842-1899) was a Norwegian mathematician who first introduced the concept of continuous symmetry.

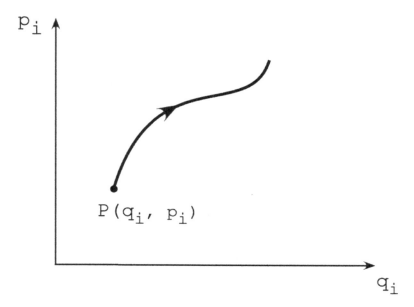

Figure 1.7   Phase space for a system described by the point P of coordinates q_i and p_i. The continuous line represents the time evolution of the system.

2 The Crisis of Classical Mechanics

Classical mechanics has been extremely successful in explaining quite a number of phenomena involving the motion of bodies from the scale of a small object with the mass of a fraction of gram, to the scale of planets revolving around the Sun. In addition, practically all electromagnetic (e.m. from now on) phenomena were very well described by Maxwell's equations, reported here in differential form and in SI units:

$$\nabla \cdot E = \frac{\rho}{\varepsilon_0}$$
$$\nabla \cdot B = 0$$
$$\nabla \times E = -\frac{\partial B}{\partial t} \tag{2.1}$$
$$\nabla \times B = \mu_0 \left(J + \varepsilon_0 \frac{\partial E}{\partial t} \right)$$

where E and B are, respectively, the vectors representing the electric field (measured in Newton per Coulomb) and the magnetic field (measured in Tesla[1]), ρ is the charge density in Coulomb per cubic meter, $\varepsilon_0 = 8.8541878176 \times 10^{-12}$ C/Vm is the vacuum electric permittivity, $\mu_0 = 4\pi \times 10^{-7}$ Vs/Am is the vacuum magnetic permeability and J is the electric current (vector) in Ampere per square meter.

In order to describe all the electromagnetism, in addition to eq. 2.1 we need the Lorentz force:

$$F = qE + qv \times B \tag{2.2}$$

Newton's laws plus Maxwell's equation were what was needed to explain *almost* all the known phenomena at the beginning of the twentieth century. The success of classical mechanics was validated by the agreement with the majority of experiments conducted in many different conditions. Classical mechanics seem to agree with experiments within the experimental errors. However, with the advancement of technology, experimental data were more and more accurate and sensitive, and deviations started to appear between data and theory. In addition, new experimental areas started to generate data that were very difficult to incorporate in a classical physics description.

We now discuss a few representative problems with classical physics that showed the inadequacy of the theory.

---

[1]The Tesla is a derived unit for magnetic field strength. It is defined as follows: a particle with a charge of 1 Coulomb, traveling with the speed of 1 meter per second, will experience a force of 1 Newton (see eq. 2.2) in a magnetic field of 1 Tesla.)

DOI: 10.1201/9781003145561-2

2.1 BLACKBODY RADIATION

Heated bodies emit e.m. radiation. In particular, if we heat a body to relatively high temperature, few thousands K, our eyes will actually see the e.m. radiation emitted as visible light. A spectacular example is our Sun which has a surface temperature around 6,000K and it appears bright and yellowish. But we also experience a sensation of heat when we expose our skin directly to the Sun: this means that the Sun emits not just at one frequency but it emits with a spectrum.

Why heated body emits e.m. radiation? A simple picture would be the following: heat is associated with random motion of atoms or molecules which are made of complex arrangements of electrically charged particles. Maxwell's equations 2.1 tell us that if we shake a charged particle it will emit e.m. radiation. The faster we shake the charged particles, then higher the frequency of the radiation emitted. Problems started to arise at the beginning of the twentieth century when, using this simple picture, physicists tried to model the spectrum of this radiation which seemed to be connected with the physics of absorption and emission of e.m. waves by various materials.

If we want to understand the physics of the absorption/emission, it is useful to idealize the process. To this end, physicists[2] have devised an idealization of the absorption/emission by introducing the concept of *blackbody*. An ideal blackbody is a material able to absorb all possible incident radiation. This means that no radiation is reflected back from a blackbody at all frequencies, or wavelengths, and all incidence angles.

Physicists at the end of the nineteenth century and beginning of twentieth century started to receive good experimental data about blackbody radiation. In particular, the data was clearly showing that the power emitted by a blackbody, per unit frequency, unit area and unit solid angle[3], is a function of only the wavelength and the temperature $B = B(\lambda, T)$ (not to be confused with the magnetic field B in Maxwell's equations). The challenge was to find a mathematical expression, based on a physical theory, able to properly describe the blackbody radiation spectrum.

In fig. 2.2, we see a modern measurement of the cosmic microwave background radiation[4] showing a characteristic spectrum with a low-frequency region, usually referred as Rayleigh-Jeans region (RJ), and a high-frequency region, usually referred as Wien region and a peak region in between. We will see in the following that classical physics can model these two extreme regions but fails to account for the peak.

The first thing to do when searching for a theory, at least if you are a classical physicist, is to make a model of the physics. Physicists therefore devised a simple model of a blackbody (see fig. 2.1): a cavity with inner sides covered with absorbing material, i.e. a material which *perfectly* absorbs radiation at all frequencies. If we

---

[2] Gustav Kirchhoff (1824-1887) was the first to introduce the concept of blackbody.

[3] Power emitted per unit area, unit solid angle and unit frequency, or wavelength, is termed Spectral Radiance.

[4] COBE satellite, FIRAS instrument, insert citation.

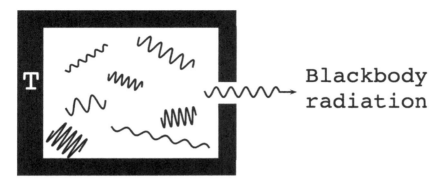

Figure 2.1 Cavity lined with perfectly absorbing material at temperature T. The radiation escaping from the small hole (bigger than the wavelengths) has a characteristic blackbody spectrum.

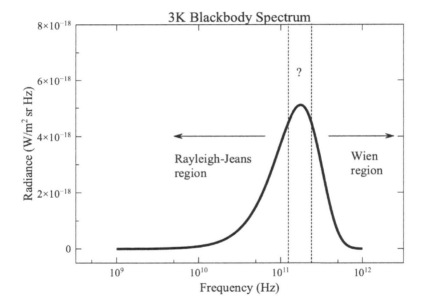

Figure 2.2 Measured spectrum (Radiance) of the Cosmic Microwave Background blackbody radiation at 2.7K.

make a small hole in this cavity at temperature T, the radiation escaping from the hole is the "blackbody" radiation whose properties we are going to discuss.

The first important distinction consists in realizing that, if $M(\lambda, T)$ is the power emitted from the surface of a blackbody at temperature T, the energy density $u(\lambda, T)$

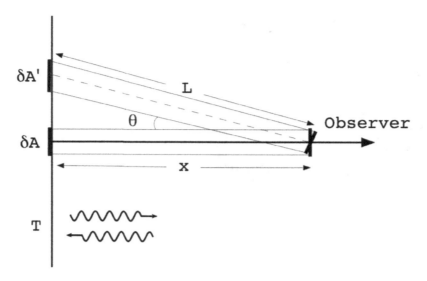

Figure 2.3 Geometry showing the relationship between energy density and power emitted by a blackbody at temperature T.

inside the cavity is:

$$u(\lambda, T) = \frac{4}{c} M(\lambda, T) \tag{2.3}$$

If we want the power emitted per unit area and unit solid angle, it can be shown that:

$$u(\lambda, T) = \frac{4}{c} M(\lambda, T) = \frac{4\pi}{c} B(\lambda, T) \tag{2.4}$$

Let's briefly justify eq. 2.3. With reference to fig. 2.3, let us consider the radiation emitted by an infinitely long wall and observed by an observer located at a distance x from the wall. Let us assume that the wall is in thermal equilibrium at temperature T. We want to compare the radiation emitted by an element δA with the energy density in the volume $V = \delta A \cdot x$, where δA is the area element and x is the distance of the observer from the wall. Remembering that power is energy per unit time, the total energy per unit wavelength within the volume V is given by:

$$\frac{dE}{d\lambda} = 2\frac{dR}{d\lambda} \cdot \tau \cdot \delta A = 2\frac{dR}{d\lambda} \cdot \frac{x}{c} \cdot \delta A \tag{2.5}$$

where factor 2 is due to the fact that, in thermal equilibrium, there is a balance of radiation emitted and absorbed thus counting twice. We assume that light propagates at speed c so that $x = c \cdot \tau$. In order to calculate the total power received by the observer, we need to average the emission from all elements seen by the observer for any angle θ from $-\pi/2$ to $\pi/2$. Because of the inclination of an angle θ, the element area radiating at an angle θ will increase geometrically by a factor $1/\cos\theta$. Another factor $1/\cos\theta$ comes from the additional length $L = x/\cos\theta$. We have that

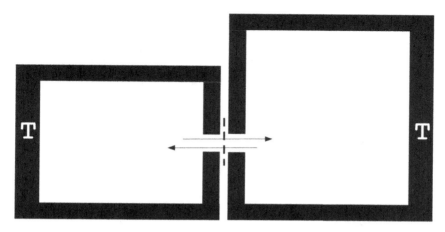

Figure 2.4 Two blackbody cavities of different shapes, but at the same temperature T, are connected through a narrow band filter allowing only radiation around the wavelength λ to pass. The energy exchanged between the two cavities must be identical; otherwise, we have a violation of the 2nd law of thermodynamics.

the power emitted, as a function of θ is equal to:

$$\frac{dE}{d\lambda}(\theta) = \frac{2}{c\cos^2\theta}\frac{dR}{d\lambda} \cdot V \tag{2.6}$$

If we now identify $B(\lambda,T) = \frac{dE}{d\lambda}$ (spectral irradiance), $u(\lambda,T) = V\frac{dR}{d\lambda}$ (spectral energy density), and we average the cosine, $\langle\cos^2\theta\rangle = 1/2$, we recover exactly eq. 2.3.

Simple thermodynamic considerations give us important information about the function $B(\lambda,T)$. First, the function $B(\lambda,T)$ depends only on the thermodynamic temperature of the emitter, i.e. the inner lining of the blackbody cavity. In fig. 2.4, two isolated blackbody cavities at the same temperature T but of different shapes and inner volume, are connected through a narrow band filter allowing to pass only radiation around a wavelength λ. Let us indicate with B_{12} the power from the left cavity through the filter into the right cavity, and with B_{21} the power from the right cavity through the filter into the left cavity. If, for example, $B_{12} > B_{21}$ then we would have a net transfer of energy from the left cavity to the right cavity resulting in the cooling of the left cavity and heating of the right cavity because they are isolated. This process would violate Lord Kelvin's statement of the second law according to which *in an isolated system it is not possible to transfer heat from one body to another at higher temperature*. It follows that $B_{12} = B_{21}$ which implies that the power emitted does not depend on internal volume. Similarly, it can be shown that the power emitted is isotropic.

We can use thermodynamics and some kinetic theory of gas to get more information about the function $B(\lambda,T)$. In particular, we want to find as much as we can, using classical physics, about the shape of the function $B(\lambda,T)$.

$$v_x dt$$

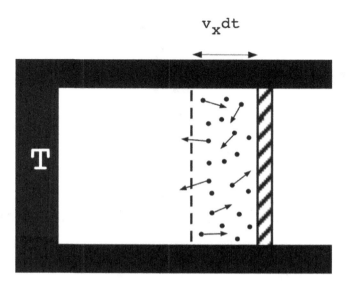

Figure 2.5 Ideal gas in an isolated container with walls in thermal equilibrium at temperature T and frictionless piston.

2.2 RADIATION PRESSURE

In order to proceed with finding the function describing the spectrum of the black-body radiation, we need to study another ideal situation: radiation inside an isolated piston, in complete analogy with the study of ideal gas in classical thermodynamics (see fig. 2.5). For a classical discussion starting from Maxwell's equations, see Longair [25]. We will use here an alternative treatment based on the kinetic theory of gases: we consider radiation inside the cavity as composed by noninteracting particles of light (photons). Following Feynman (Vol I, 39-1) [16], let's evaluate the pressure exerted from a perfect mono-atomic gas enclosed in a cylinder with a frictionless piston (fig. 2.5).

We want to evaluate the pressure P exerted by the gas at temperature T on the frictionless piston. The pressure $P = F/A$ is the force F exerted by the particles hitting the piston divided by the area A of the piston. The total force will be the sum of the force exerted by each individual particle which, in turn, is the variation of momentum dp per unit time t. Since the piston is bound to move only horizontally along the x direction, only the x component of the momentum p_x will contribute to the force. The momentum change due to a particle hitting the piston is $2\,dp_x = 2mv_x$. If we have N particles in the inner volume V, only those in the volume $V = Av_x dt$ will hit the piston during the time dt. The total number of particles hitting the piston will be the volume V times the number of atoms per unit volume $n = N/V$ and the pressure is:

$$P = \frac{2}{A}\frac{dp}{dt} = 2nmv_x^2 = nm\langle v_x^2\rangle \tag{2.7}$$

where the last equality takes into account that not all the particles are moving toward the piston. On average, only half will move toward the piston and half away from it.

In a totally disordered system, like our ideal gas, the particles move in all direction with equal probability. This means that $\langle v_x \rangle^2 = \langle v_y \rangle^2 = \langle v_z \rangle^2$ and a completely random motion requires that:

$$\langle v_x^2 \rangle = \frac{1}{3}(\langle v_x^2 \rangle + \langle v_y^2 \rangle + \langle v_z^2 \rangle) = \frac{\langle v^2 \rangle}{3} \qquad (2.8)$$

which inserted in eq. 2.7 gives:

$$P = \frac{2}{3}n\langle \frac{mv^2}{2} \rangle \qquad (2.9)$$

In a perfect gas made of monoatomic noninteracting particles the internal energy U is just the kinetic energy and therefore:

$$P = \frac{2}{3}u \qquad (2.10)$$

where $u = U/V$ is the energy density.

What happens if we now treat the radiation inside the cavity as a collection of massless particles moving at the speed of light c? If we repeat the above reasoning, we need to express the momentum of each particle of radiation with $p = E/c$. Let's rewrite eq. 2.9 as:

$$P_\gamma = \frac{1}{3}n\langle pv \rangle = \frac{1}{3}n\langle pc \rangle \qquad (2.11)$$

where P_γ is the radiation pressure. We know from special relativity that the energy of photons is $E = pc$, which when inserted into 2.11 gives:

$$P_\gamma = \frac{1}{3}u. \qquad (2.12)$$

The result 2.12 is purely classical and can be derived using Maxwell's equations[5] as mentioned at the beginning of this section.

We can write eq. 2.12 by remembering that $u = U/V$:

$$P_\gamma V = \frac{1}{3}U = (\gamma - 1)U \qquad (2.13)$$

where the constant $\gamma = 4/3$ is the so-called *adiabatic index* of photons. Notice that the subscript γ of the radiation pressure P_γ indicates instead the radiation pressure of photons which are historically indicated also as gamma rays.

---

[5] In fact, it is well known that Maxwell's equations are relativistically invariant.

2.3 STEFAN-BOLTZMANN LAW

Armed with the results of the previous section, it is now relatively straightforward to obtain the integrated power emitted by a blackbody at temperature T. Based on the measurements of two physicists, Dulong[6] and Petit[7], in 1879 Josef Stefan[8] proposed a law for the total power emitted by a blackbody. Shortly after, Boltzmann[9] derived Stefan's law using classical thermodynamics.

We start with the first principle of thermodynamics:

$$dU = \delta Q + \delta W \tag{2.14}$$

where U, Q and W are, respectively, the internal energy, the heat exchanged with the system and the work done on the system. The symbol δ indicates that both Q and W are not functions of state although their sum is. In the case of reversible changes, then we can use the fundamental thermodynamic relation to write the first law in terms of state functions:

$$dU = TdS - PdV \tag{2.15}$$

if we derive eq. 2.15 with respect to the volume V, keeping the temperature T constant, we have:

$$\left(\frac{\partial U}{\partial V}\right)_T = T\left(\frac{\partial S}{\partial V}\right)_T - P \tag{2.16}$$

Since we are looking for a relationship between energy U and temperature T, we want to eliminate the entropy derivative from eq. 2.16. This is easily achieved by using one of the Maxwell's relations, namely:

$$\left(\frac{\partial P}{\partial T}\right)_V = \left(\frac{\partial S}{\partial V}\right)_T \tag{2.17}$$

It follows that:

$$\left(\frac{\partial U}{\partial V}\right)_T = T\left(\frac{\partial P}{\partial T}\right)_V - P \tag{2.18}$$

using $U = uV$ and eq. 2.12 we have:

$$\left(\frac{\partial(uV)}{\partial V}\right)_T = T\left(\frac{\partial \frac{u}{3}}{\partial T}\right)_V - \frac{u}{3} \tag{2.19}$$

---

[6] Pierre Louis Dulong (1785-1838) was a French physicist and chemist. Together with Alexis Petit first showed that the heat capacity of metals are inversely proportional to their masses.

[7] Alexis Therese Petit (1791-1820) was a French physicist better known for his work with Dulong on heat capacity of metals.

[8] Josef Stefan (1835-1893) was an Austrian physicist who first estimated the temperature of the Sun's surface by using his empirical law.

[9] Ludwig Boltzmann (1844-1906) was an Austrian physicist considered to be the father of classical statistical mechanics.

which simplifies into:

$$u = \frac{1}{3}T\left(\frac{\partial u}{\partial T}\right) - \frac{u}{3} \tag{2.20}$$

which, after a simple integration, gives:

$$u = \sigma T^4 \tag{2.21}$$

where σ is the *Stefan-Boltzmann constant*. Thermodynamics alone is not capable of giving the numerical value of the constant of eq. 2.21 which was first determined experimentally. The Stefan-Boltzmann constant has a numerical value of $\sigma = 5.7 \times 10^{-8} W/(m^2 K^4)$ and we will see later that it is possible to express it in terms other fundamental units. Notice that eq. 2.21 implies that the integrated spectral irradiance is:

$$\int_0^\infty B(\lambda, T)d\lambda = \sigma T^4 \tag{2.22}$$

2.4   WIEN'S DISPLACEMENT LAW

We can push a bit further the thermodynamics (and the electromagnetism) and show, as Wien did, that we can give constraints to the functional form of the blackbody radiation. We have already used the "trick" of considering the blackbody radiation inside a cavity as a thermodynamic system composed of a "gas" of radiation. We did not investigate more what we actually mean with a "gas" of radiation. Is radiation made up of non-interacting particles? Actually yes and we will see later that the photon is the quantum of electromagnetic radiation. For now, let's consider a classical gas of radiation without specifying much about the individual particles. Let's consider again a gas of radiation inside an isolated box with reflecting walls and let us operate an adiabatic expansion. Let's rewrite the first law in the standard form:

$$dQ = dU + pdV \tag{2.23}$$

Following Longair [25], we know that an adiabatic expansion is defined by $dQ = 0$, i.e. no exchange of heat between the system and the environment during the expansion. We have also shown that $U = uV$ and $p = u/3$. Therefore:

$$d(uV) + \frac{1}{3}udV = 0$$
$$Vdu + udV + \frac{1}{3}udV = 0 \tag{2.24}$$
$$\frac{du}{u} = -\frac{4}{3}\frac{dV}{V}$$

This last equation integrates to:

$$u = \text{const} \times V^{-4/3} \tag{2.25}$$

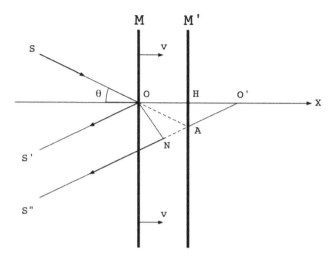

Figure 2.6   Geometry showing the Doppler shift of radiation reflected off a receding mirror.

Using eq. 2.22, we have:

$$\sigma T^4 = \text{const} \times V^{-4/3}$$
$$TV^{1/3} = \text{const} \tag{2.26}$$

If we now consider a spherical volume $V = \frac{4}{3}\pi R^3$, eq. 2.26 implies that:

$$T \propto \frac{1}{R} \tag{2.27}$$

This last result is extremely powerful and interesting. Eq. 2.27 tells us that, if we have an expanding box containing blackbody radiation at temperature T, the temperature will increase or decrease if we compress or expand the box[10].

We now want to find what happens to the wavelength λ of the blackbody radiation as the box expands. Having assumed that the inner surface of our expanding box is perfectly reflecting, we need to study what happens to the wavelength of the radiation as is reflected off a slowly receding mirror.

In fig. 2.6 a source of monochromatic radiation S is illuminating a mirror M receding at a velocity v (along the x-xis) away from the source. The mirror is moving with a velocity v such that, after one period T of the wave, it will reach the position

---

[10] An extreme consequence of this result would imply that the Cosmic Backgrund Radiation, i.e. the radiation emitted by the primordial Universe at the time of last scattering, has today a temperature much lower ($\sim 2.73K$) than the emission temperature ($\sim 3,000K$) about 13.7 billion years ago, few hundred thousands years after the Big Bang. This cooling is a direct effect of the isotropic expansion of the Universe.

M'. We assume that $v \ll c$, where c is the speed of light so as to neglect relativistic effects. The radiation is making an angle θ with respect to the direction x of the velocity v. In the reference frame of the source S or the observers S' and S'', the wavelength of the radiation is Doppler shifted by the receding mirror. First, let's discuss the geometry where the source is on the x-axis, i.e. the angle θ is equal to zero. It is well known that, *a receding source* generates a Doppler shift equal to $\Delta \lambda = \lambda \frac{v}{c}$ which is the amount of red shift seen by an observer sitting on the receding mirror. The receding mirror is now the source of the reflected radiation and it will generate an additional Doppler shift when observed by the observer in S. The total Doppler shift will be:

$$\Delta \lambda = 2\lambda \frac{v}{c} \tag{2.28}$$

the wavelength stretch $\Delta \lambda$ being equal to twice the segment $\overline{OH} = \overline{HO'}$. A similar argument applies when the radiation is arriving tilted with an angle θ, as shown in fig. 2.6. Now the wavelength stretch is equal to the sum of the two segments $\overline{OA}$ and $\overline{AN}$, where the segment $\overline{ON}$ is perpendicular to the line $\overline{O'S''}$. Using simple geometry on the triangles $\triangle OO'N$ and $\triangle OO'A$ we have:

$$\Delta \lambda = \overline{OA} + \overline{AN} = \overline{O'A} + \overline{AN} = \overline{OO'} \cos \theta \tag{2.29}$$

But we know that $\overline{OO'} = 2\overline{OH} = 2vT$, where T is the period of the wave. So, the generalization of formula 2.28 for arbitrary angles θ is therefore:

$$\Delta \lambda = 2\lambda \frac{v}{c} \cos \theta \tag{2.30}$$

Having studied a single reflection from a receding mirror, following Longair again [25] let us now study the case of multiple reflections in a spherical cavity (see fig. 2.7). During a (slowly) expansion of the cavity from radius r to radius $r + dr$, the radiation will be subject to multiple reflections keeping the angle θ constant. The time between two successive reflections will be $\overline{AB}/c = 2r \cos \theta/c$. In the time $dt = dr/v$, the sphere will have increased its radius by dr. In the same time interval dt the number of reflections will be the ratio between dt (the time of the expansion) and the time between reflections $\overline{AB}/c = 2r \cos \theta/c$. This ratio is $cdt/(2r \cos \theta)$. It follows immediately that the change in wavelength is the product of the number of reflections in the time dt times the Doppler shift due to one reflection as in eq. 2.30:

$$d\lambda = \left(2\lambda \frac{v}{c} \cos \theta\right) \left(\frac{cdt}{2r \cos \theta}\right) \tag{2.31}$$

$$\frac{d\lambda}{\lambda} = \frac{vdt}{r} = \frac{dr}{r}$$

which integrated gives:

$$\lambda \propto r \tag{2.32}$$

We can finally put together eq. 2.27 and eq. 2.32 to give:

$$T \propto \lambda^{-1}$$
$$\lambda T = \text{const} \tag{2.33}$$

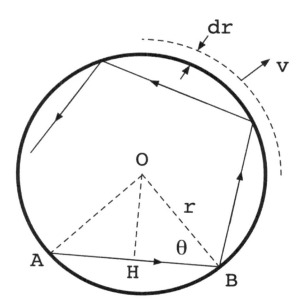

Figure 2.7 Geometry showing the Doppler shift of radiation reflected off an expanding spherical cavity.

This last equation is often referred as *Wien's Displacement Law*. The value of the constant in eq. 2.33, usually indicated with b, has to be determined experimentally. If we use the peak wavelength λ_{max}, then eq. 2.33 can be written as:

$$\lambda_{max}T = 2.898 \cdot 10^{-3} \text{ m} \cdot \text{K} \tag{2.34}$$

Eq. 2.34 has been experimentally verified to a good accuracy and we will see in the next chapter that it can be derived easily from the Planck's black body spectral function.

It is interesting to briefly discuss if a law similar to eq. 2.34 can be written for frequency instead of wavelength. The direct substitution of $v = c/\lambda$ generates a formula which is not in agreement with experiments. This is telling us that the function we are after is not a simple function of an independent variable. If we assume that the function describing the spectral radiance of a blackbody is a *distribution function*, then we obtain the correct formula for the Wien's displacement law expressed in frequencies instead of wavelengths [9]:

$$\frac{v_{max}}{T} = 5.879 \cdot 10^{10} \text{ K}^{-1}\text{s}^{-1} \tag{2.35}$$

We will justify eq. 2.35 after we have obtained Planck's formula for the spectral radiance.

Let's make a brief recap of our journey toward the determination, using classical physics, of the shape of the spectral radiance $B(\lambda, T)$ of a blackbody at temperature

T. Using Maxwell's equations and classical thermodynamics, we have seen that we can derive the Stefan-Boltzmann law and the Wien's displacement law. It is possible, as Wien did, to go a bit further and give some constraints to the shape of the spectral radiance of a blackbody. Following Longair [25], let's consider a box containing blackbody radiation in equilibrium with the perfectly absorbing (and emitting) enclosure at temperature T_1. Let the box adiabatically expand in such a way that the radiation has time to thermalize with the enclosure at the final temperature T_2. Since we know that $T \propto 1/\lambda$ the radiation spectrum will keep its blackbody form during the expansion when λ is stretched.

Let us now consider the wavelength interval between λ_1 and $\lambda_1 + d\lambda_1$ with corresponding energy density $u = B(\lambda_1)d\lambda_1$. Using eq. 2.22, we have:

$$\frac{B(\lambda_1)d\lambda_1}{B(\lambda_2)d\lambda_2} = \frac{T_1^4}{T_2^4} \tag{2.36}$$

Using Wien's displacement law and $d\lambda_1 = T_2/T_1 d\lambda_2$:

$$\frac{B(\lambda_1)}{T_1^5} = \frac{B(\lambda_2)}{T_2^5} = \text{const} \tag{2.37}$$

Eq. 2.33 tells us that the combination λT is constant and therefore eq. 2.37 becomes:

$$B(\lambda)\lambda^5 = \text{const} = f(\lambda T)$$
$$B(\lambda)d\lambda = \frac{1}{\lambda^5}f(\lambda T)d\lambda \tag{2.38}$$

which is Wien's displacement law setting the form of the blackbody radiation spectrum. If we want to express eq. 2.38 in terms of frequencies, we first notice that we must have:

$$B(\lambda)d\lambda = B(v)dv$$
$$\lambda = \frac{c}{v} \tag{2.39}$$
$$d\lambda = -\frac{c}{v^2}dv$$

Incidentally, we notice that eq. 2.39 tells us that the function B is a *distribution function*. It is now easy to derive Wien's displacement law in terms of frequencies:

$$B(v)dv = v^3 f(\frac{v}{T})dv \tag{2.40}$$

Eqs. 2.38 and 2.40 show the power of thermodynamics and Maxwell's equations to derive constraints on the functional form of the blackbody radiation. Wien went a bit further [45] and empirically proposed a particular function for $f(\lambda T)$:

$$f(\lambda T) = \frac{C_1}{\lambda^5}e^{\frac{C_2}{\lambda T}} \tag{2.41}$$

where C_1 and C_2 are two constants to be determined by experiments[11]. Experiments showed that Wien's approximation 2.41 well-represented short wavelengths but failed to account the long wavelength regime. This formula is quite remarkable because it predicts that the shape of the blackbody spectrum does not change with the temperature T as it is effectively confirmed by experimental data.

2.5 THE EQUIPARTITION THEOREM AND THE RAYLEIGH-JEANS LAW

We can try to attack the problem of finding the shape of the blackbody radiation spectrum by inserting a bit more physics. Let's assume that we want to study the electromagnetic radiation in thermal equilibrium with a box with perfectly reflecting walls as was originally done by the two physicists Rayleigh and Jeans. Inside the box, there will be a collection of standing waves and our task is to try to calculate the number of modes. Once we know how many modes per frequency we have, we can calculate what energy is associated with each mode and the blackbody radiation spectrum is obtained once we finally calculate the energy density per unit frequency (or wavelength). The calculation is therefore done in two steps: counting the modes per frequency interval and assigning the proper energy to each mode.

The calculation of number of modes inside a cavity is simplified if we assume that the cavity is a cube of side L. We know that standing waves occur when, for each wavelength λ, there is an integer number of half-wave cycles between two reflections. In the simplest case of a standing wave along one of the sides of the cube, the standing wave condition is:

$$n = \frac{L}{\frac{\lambda}{2}} = \frac{2L}{\lambda} \tag{2.42}$$

where $n = 1, 2, \ldots$ is an integer. In fig. 2.8 the first three standing waves, corresponding to $n = 1, 2$ and 3 are shown.

We need to take into account that there are two independent polarizations of the wave with the same index n: this factor 2 in the computation of the total number of modes will be inserted at the end of the calculation.

We can rewrite eq. 2.42 in a more convenient way by using the wavenumber $k = 2\pi/\lambda$ and the relationship between frequency and wavelength $v = c/\lambda$:

$$k = \frac{\pi n}{L} \tag{2.43}$$

and squaring eq. 2.43 we have:

$$k^2 = \pi^2 \left(\frac{n}{L}\right)^2 \tag{2.44}$$

---

[11]We will see that Planck gave the values of the two constants in terms of the Boltzmann constant k, the speed of light c and his new constant h, $C_1 = 2\pi h c^2$ and $C_2 = -hc/k$

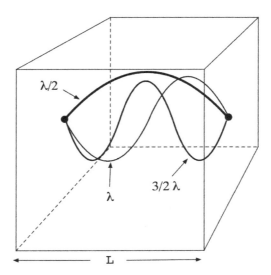

Figure 2.8   First three standing wave modes inside a cubic cavity with reflecting walls of side L.

If we consider the three independent directions along the sides of the cubic box, eq. 2.44 can be generalized to:

$$k^2 = \pi^2 \left[\left(\frac{n_x}{L}\right)^2 + \left(\frac{n_y}{L}\right)^2 + \left(\frac{n_z}{L}\right)^2 \right] \tag{2.45}$$

We can now rewrite eq. 2.45 as:

$$n^2 = n_x^2 + n_y^2 + n_z^2 = \frac{L^2 k^2}{\pi^2} \tag{2.46}$$

where n is a vector of components (n_x, n_y, n_z). In order for eq. 2.46 to properly represent standing waves, the indexes (n_x, n_y, n_z) must be positive integers. Notice that eq. 2.46 represents the equation of a sphere in the indexes space. In order to find the total number of standing waves we have to just calculate the volume of such a sphere, providing that we limit ourselves to the fraction of the sphere where the indexes (n_x, n_y, n_z) are positive. Such a fraction is exactly $1/8$ and therefore we have that the total number of standing waves N is:

$$N = 2\frac{1}{8}\frac{4}{3}\pi n^3 = \frac{\pi}{3}n^3 \tag{2.47}$$

where the factor 2 in front takes into account of the two polarizations per standing wave as discussed above. Using eq. 2.46 and the definition of wavenumber we have:

$$N = \frac{\pi}{3}\left(\frac{L^2 k^2}{\pi^2}\right)^{\frac{3}{2}} = \frac{\pi}{3}\left(\frac{Lk}{\pi}\right)^3 = \frac{8\pi L^3}{\lambda^3} \tag{2.48}$$

This last equation gives the total number of modes summed over all the possible wavelengths. If we want to find the number of modes per unit wavelength we have to differentiate:

$$\frac{dN}{d\lambda} = -\frac{8\pi L^3}{\lambda^4} \tag{2.49}$$

we should not be surprised about the minus sign in eq. 2.49 because the number of modes decreases with increasing the wavelength. If we want the number of modes per unit wavelength per unit volume we just divide eq. 2.49 by the volume of the cavity $V = L^3$:

$$\frac{1}{V}\frac{dN}{d\lambda} = -\frac{8\pi}{\lambda^4} \tag{2.50}$$

In order to find the energy density $u(\lambda, T)$ inside the cavity we need to assign the corresponding energy to each mode. Before progressing to find the energy density we are left to discuss two issues: 1) how many degrees of freedom one mode of an electromagnetic wave inside a cavity has?; 2) how much energy is assigned to each mode?

We have briefly studied in chapter 1 the harmonic oscillator and we have seen that it has two degrees of freedom: one for the kinetic energy and one for the potential energy. We could simply say that we have thermal equilibrium between the radiation inside the cavity and the walls at temperature T. We can assume that each mode of radiation in the cavity is coupled to a specific oscillating charged particle in the wall. Requiring thermal equilibrium is equivalent to require that the energy of each oscillating charged particle is in equilibrium with the corresponding mode in the cavity and therefore each e.m. mode has exactly two DOFs. Although correct, this argument is just qualitative and a bit more rigorous treatment is desirable.

Let's start with Maxwell's equations in vacuum, i.e. $\rho = J = 0$. We can make the third and fourth equations symmetrical if we remember that $\mu_0 \varepsilon_0 = 1/c^2$:

$$\nabla \cdot E = 0$$
$$\nabla \cdot B = 0$$
$$\nabla \times E = -\frac{1}{c}\frac{\partial B}{\partial t} \tag{2.51}$$
$$\nabla \times B = \frac{1}{c}\frac{\partial E}{\partial t}$$

It can be shown[12] that we can combine eqs. 2.51 into:

$$\nabla^2 E - \frac{1}{c^2}\frac{\partial^2 E}{\partial t^2} = 0 \tag{2.52}$$

This equation has a general solution $E = E(\omega t - \vec{k} \cdot \vec{x})$ of the form of a traveling wave of arbitrary shape of the argument $(\omega t - \vec{k} \cdot \vec{x})$, where $\omega = 2\pi\nu$,

---

[12]Use the vector identity $\nabla \times (\nabla \times a) = \nabla(\nabla \cdot a) - \nabla^2 a$.

$\vec{k} = (k_x, k_y, k_z)$ is the wavenumber vector and $\vec{x} = (x, y, z)$ is the vector of the coordinates.

By fixing the boundary conditions at the walls of the cavity, i.e. $E(0) = E(L) = 0$, the field $E(\vec{x}, t)$ can be written as the summation of the product of two functions, one $f_m(t)$ depending only on the time t, and one $g_m(\vec{x})$ depending only on the coordinate vector $\vec{x}$:

$$E(x, t) = \sum_m f_m(t) \cdot g_m(\vec{x}) \qquad (2.53)$$

where the functions $g_m(x)$ are the so-called *normal modes* and the index m accounts for the progressive mode number. Notice that there is an infinite number of terms in 2.53. The boundary conditions inside the cavity impose that:

$$\begin{aligned} \nabla^2 g_m(\vec{x}) + k_m^2 g_m(\vec{x}) \\ \nabla \cdot g_m(\vec{x}) = 0 \\ \vec{n} \times g_m(\vec{x}) = 0 \end{aligned} \qquad (2.54)$$

The normal modes satisfy the *orthonormality condition*:

$$\int g_m(\vec{x}) g_n(\vec{x}) d^3 x = \delta_{m,n} \qquad (2.55)$$

thus justify the name "normal modes". If we now substitute the expression 2.53 into the wave equation 2.52 we find an equation for the time-dependent functions $f_m(t)$:

$$\sum_m \frac{d^2 f_m}{dt^2} + c^2 k_m^2 f_m(t) = 0 \qquad (2.56)$$

The functions $f_m(t)$ are linearly independent and therefore eq. 2.56 must be valid for each element of the summation:

$$\frac{d^2 f_m}{dt^2} + c^2 k_m^2 f_m(t) = 0 \qquad (2.57)$$

If we rewrite eq. 2.56 with Newton's notation $\ddot{f}_m + c^2 k_m^2 f_m$ we notice a striking similarity with eq. 1.29 $\ddot{q} + \omega^2 q = 0$ describing the simple harmonic oscillator. This means that **the modes of the electric field inside a cavity are mathematically and physically equivalent to a collection of independent simple harmonic oscillators with frequency** $\omega_m = c k_m$. In particular, we have found that each standing wave mode contribute 2 DOFs just like a simple one-dimensional harmonic oscillator. It is important not to identify these two DOFs with the two independent polarization of each mode: there are 2 DOFs per mode per polarization[13].

---

[13] We have determined the number of DOFs by studying the electric field of the e.m. inside the cavity. The reader might wonder if there are two additional DOFs associated with the magnetic field. This is not the case because the Maxwell-Faraday law of induction $\nabla \times E = -\partial B / \partial t$ states that a time-varying magnetic field induces a spatially varying nonconservative electric field and vice versa therefore somehow "sharing" the DOFs

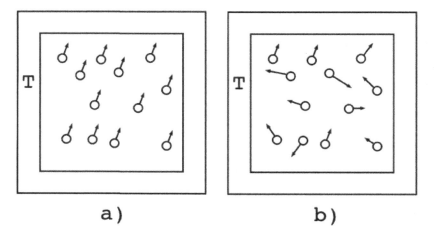

Figure 2.9   Perfect gas thermalizing from a) to b).

Before finally determining the Rayleigh-Jeans formula, we need to assign the energy to each of the DOFs of the modes. The *equipartition theorem* does exactly this: in a system in thermal equilibrium at a temperature T, the energy is shared equally among all energetically accessible DOFs of the system. This is not particularly surprising because the theorem assess that the system is maximizing its entropy by distributing the energy over all possible accessible modes. The equipartition theorem tells us more: each quadratic DOF, i.e. each DOF that appears quadratically in the energy expression, will possess an energy $\frac{1}{2}k_BT$, where k_B is the Boltzmann constant. A proof of the equipartition theorem requires statistical mechanics and we refer the reader to specialized books (see, for example Mandl[26]). Here we show a simple example calculation of the equipartition in the case of a box containing N free particles as in fig. 2.9. Panel a) of fig. 2.9 shows a collection of classical particles placed in various position inside a box whose walls are at temperature T. All the particles have exactly the same velocity vectors. As a result of repeated collisions with the walls, the particles will spread out and assume various velocity vectors with different magnitude and direction and, after some time, the particles will look something like fig. 2.9 panel b) where a spread in position and velocity is now evident.

Statistical mechanics tell us that the speed of particles of mass m, non interacting among themselves but in thermal contact with surrounding walls at temperature T, obey the so-called Maxwell-Boltzmann distribution:

$$f(v) = 4\pi \left(\frac{m}{2\pi kT}\right)^{3/2} v^2 e^{\left(\frac{-mv^2}{2kT}\right)} \tag{2.58}$$

This function is related to the probability of finding particles with a certain velocity. The problem is that, mathematically, the probability of finding a particle with *exactly* a certain velocity is zero. In this sense $f(v)$ is a *generalized function* and must be interpreted in the following way: $f(v)dv$ is the probability to find the particle in

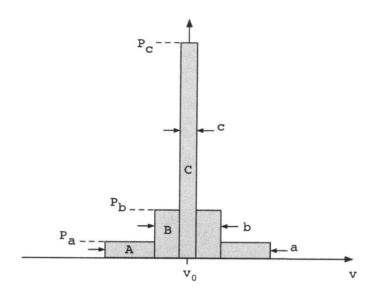

Figure 2.10   Dirac delta function construction. All the boxes have the same area equal to 1.

the velocity interval between v and $v + dv$. In order to make sure that the function $f(v)$ represents a probability, it must be normalized in such a way that:

$$\int_{-\infty}^{+\infty} f(v)dv = 1 \qquad (2.59)$$

Given the distribution function $f(v)$, we can calculate averages of quantities. If we want to calculate, for example, the average kinetic energy of the particles of the gas in the condition a) of fig. 2.9, we can simply state that it is $\frac{1}{2}mv_0^2$, having all particles the same mass m and the same velocity v_0. In this trivial case, the distribution function of the velocities is a special function that is 0 everywhere except when the speed is exactly v_0. What is the value of $f(v)$ when $v = v_0$?

We can try to guess $f(v_0)$ by assuming that the particles have all *almost* the same velocity as depicted in fig. 2.10. The box A of width a assigns the same probability P_a to all particles having velocities comprised between $v_0 - a/2$ and $v_0 + a/2$. If we want to be more accurate, we can build a narrower box B of width b assigning a probability P_b to all particles with velocities between $v_0 - b/2$ and $v_0 + b/2$. In a similar way, we can make an even narrower box C of width c assigning a probability P_c to all particles with velocities between $v_0 - c/2$ and $v_0 + c/2$. All these boxes obey the same probability normalization condition 2.59 which, in these cases, is equivalent to asking that all the boxes have area equal to 1. If we keep narrowing boxes keeping the area constant, we notice that the value of the probability function keeps increasing. Obviously, when we require to assign exactly the value v_0 with the interval around it going to zero, the probability goes to infinity, but it does so by keeping *the area equal to 1*, i.e. satisfying the normalization condition of the probability. This strange function, which we now know it is not a proper function but rather a generalized function, is indicated as *delta function* (δ function) and was first introduced by the British physicist P.A.M. Dirac[14]. The normalization condition for the delta function is therefore:

$$\int_{\infty}^{\infty} \delta(v)dv = 1 \tag{2.60}$$

In a slightly more mathematical terms, we can make a more general statement and say that the delta function is the limit of a delta convergent sequence. This means that other mathematical functions, like the simple rectangles considered above, can be used to define the delta function as a limit for $v \to 0$ of a sequence. For example, we can use Gaussians of area 1 such that:

$$\lim_{v \to 0} \frac{1}{\sqrt{2\pi\alpha}} e^{-\frac{v^2}{4\alpha}} \to \delta(v) \tag{2.61}$$

An equivalent definition of the delta function is:

$$\int_{-\infty}^{+\infty} f(v)\delta(v) = f(0) \tag{2.62}$$

or more general:

$$\int_{-\infty}^{+\infty} f(v)\delta(v - v_0) = f(v_0) \tag{2.63}$$

Notice that the delta function is defined in terms of integration and when operated on a function it returns a scalar number.

We now go back to express the average kinetic energy of the gas in the condition a) of fig. 2.9. Having introduced the special distribution function $\delta(v)$ describing the velocities, the average kinetic energy per particle is:

$$\langle T \rangle = \int_{-\infty}^{+\infty} \frac{1}{2}mv^2 \delta(v - v_0)dv = \frac{1}{2}mv_0^2 \tag{2.64}$$

More interesting is the case of condition b) of fig. 2.9. We want to show that the calculation of the average kinetic energy agrees with the Equipartition Theorem. Using Maxwell-Boltzmann distribution of velocities 2.58 and considering only the magnitude of the velocities, we have:

$$\langle T \rangle = \int_0^\infty \frac{1}{2}mv^2 f(v)dv = \frac{4\pi m}{2} \int_0^\infty v^4 e^{-\frac{mv^2}{2kT}} \tag{2.65}$$

The integral in eq. 2.65 can be evaluated by substituting $x = v$, $a = \frac{m}{2kT}$ and using a known result:

$$\int_0^\infty x^{2n} e^{-ax^2} = \frac{(2n-1)!!}{2^{n+1}a^n} \left(\frac{\pi}{a}\right)^{\frac{1}{2}} \tag{2.66}$$

---

[14]The δ function is also called Dirac's δ.

where $n!! = n(n-2)(n-4)...$ and so on. It follows:

$$\langle T \rangle = 3\pi m \left(\frac{m}{2\pi kT} \right)^{3/2} \left(\frac{2\pi kT}{m} \right)^{1/2} = \frac{3}{2}kT \qquad (2.67)$$

We have just proved that the average kinetic energy of a particle of a non-interacting gas, having just 3 translational degrees of freedom, is $\frac{3}{2}kT$ in agreement with the Equipartition Theorem[15].

We are finally ready to express the Rayleigh-Jeans approximation for the blackbody radiation inside a cavity. We have calculated in eq. 2.50 the number of modes per unit wavelength and we have seen that each mode has two degrees of freedom. We can finally write that the energy density is:

$$u(\lambda, T)d\lambda = \frac{8\pi}{\lambda^4}kTd\lambda \qquad (2.68)$$

notice that we dropped the minus sign which was only intended to show the inverse relationship between number of modes and wavelength. We can express the energy density in terms of frequency by imposing $u(\lambda, T)d\lambda = u(v, T)dv$. Inserting $\lambda = c/v$ and $d\lambda = -c/v^2 dv$ into eq. 2.68, we find:

$$u(v, T)dv = \frac{8\pi v^2}{c^3}kTdv \qquad (2.69)$$

Remember that eq. 2.69 gives the energy density of the blackbody radiation inside the cavity. If we want to know the radiation emitted by the blackbody $B(\lambda, T)$ we need to use eq. 2.3:

$$B(\lambda, T)d\lambda = \frac{c}{4}u(\lambda, T)d\lambda = \frac{2\pi c}{\lambda^4}kTd\lambda$$
$$B(v, T)dv = \frac{2\pi v^2}{c^2}kTdv \qquad (2.70)$$

Eqs. 2.70 are known as *Rayleigh-Jeans* approximation to the radiation emitted by a blackbody.

2.6 THE ULTRAVIOLET CATASTROPHE

We have reached the end of our journey whose task was to determine the functional form of the spectral dependence of the blackbody radiation. On the short wavelengths side (high frequency) we have the Wien's approximation expressed by eq. 2.41 which was proposed empirically by Wien. On the long wavelength side (low frequency) we have instead the Rayleigh-Jeans approximation in eq. 2.70 which has been derived using a physical model and using the equipartition of energy. The main problems of

---

[15]The general derivation of the Equipartition Theorem is beyond the scope of this book and can be found in any good book on classical statistical mechanics.

the only physical model - the Rayleigh-Jeans - were twofold (somehow related): 1) it did not account for the peak and 2) it predicted that the energy density inside a cavity will diverge since:

$$\int_0^\infty B(v,T)dv = \int_0^\infty \frac{2\pi v^2}{c^2} kT dv \to \infty \tag{2.71}$$

which is clearly absurd and blatantly in violation of experimental facts: we do not observe emission of gamma rays of higher and higher energy coming out of an oven. This problem is referred in the literature as *the ultraviolet catastrophe* and posed a huge problem to the physicists at the end of the nineteenth century. Physicists had to think about what assumptions are not justified and needed to be changed. The physics of oscillators coupled to e.m. waves was clearly solid: Maxwell's equation are still valid today. Therefore something else must be wrong: the equipartition of energy was the next suspects.

Section II

Quantum Mechanics

3 From Classical to Quantum Physics

There is a story, not verified[1], that Lord Kelvin has claimed, quite boldly around the year 1890, that all the physics was practically known and what was left to do was to refine the values of the measured fundamental constants. This statement was enunciated to celebrate the success of what we call today "classical mechanics" and it is nowhere to be found in any lectures or writings of Kelvin[2].

In spite of the authenticity or less of the quote, we can be almost certain that these feelings of "completeness" of physics were around toward the end of nineteenth among all the influential physicists of the time. Newtonian mechanics and Maxwell's theory of e.m. phenomena were able to explain most of the experimental facts known at the time. The development of technology and more accurate measurement systems, however, brought new data to the scientific community of the time. Some of these new measurements were puzzling. As we have seen previously, there were experimental but also theoretical issues related to the blackbody radiation whose spectrum seemed to be defying all the attempts to be derived using classical physics. Wien's approximation worked well at high frequency (short wavelengths) while Rayleigh-Jeans approximation worked well at low frequencies (long wavelengths). Both approximations failed to describe the peak region.

In the year 1900, in a lecture delivered at the Royal Institution of Great Britain [22], Kelvin pointed out that there were *two clouds* obscuring the *beauty and clearness of the dynamical theory which asserts heat and light to be modes of motion*[3]. In his own words: 1) How could earth move through an elastic solid, such as essentially is the luminiferous ether? 2) the Maxwell-Boltzmann doctrine regarding the partition of energy.

The answers to these two questions determined an amazing revolution in the understanding of our Universe: the theory of relativity and quantum mechanics. These two theories have passed all experimental tests and, as of today, they are still considered to be the best description of nature. We will deal only with the second cloud.

---

[1] It is well known that science textbooks show a very poor account of history of science: understandably textbooks tend to organize the material in a logical progression rather than the historical one. This book is no exception but where possible the author has tried to avoid common misconceptions.

[2] It has been suggested that this quote should be attributed instead to the American physicist Albert A. Michelson

[3] For an historical discussion pointing out common misconceptions about the clouds, see [33].

DOI: 10.1201/9781003145561-3

3.1 PLANCK'S FORMULA

We have already briefly encountered the German physicist Gustav Kirchhoff[4] when we introduced the concept of blackbody in the previous chapter. Kirchhoff studied extensively the physics of blackbody radiation and enunciated his *Kirchhoff's law of thermal radiation*: "Given an arbitrary body emitting and absorbing thermal radiation in thermodynamic equilibrium, the emissivity is equal to the absorptivity". The emissivity of a body $\varepsilon(v)$ is defined as the ratio between the radiation $I(v,T)$ emitted by the surface of a body at temperature T and the radiation $B(v,T)$ emitted by an ideal blackbody so that we can write that $I(v,T) = \varepsilon(v)B(v,T)$. Notice that, in general, the emissivity is a function of the frequency v. The absorptivity $\alpha(v)$ is the fraction of the incident radiation that is absorbed by the body. Kirchhoff's law then is simply $\alpha(v) = \varepsilon(v)$. Kirchhoff went a bit further and *postulated* that the function $B(v,T)$ is a universal function and is the same for all perfect black bodies. Wien's and Rayleigh-Jeans are approximations to this universal function whose exact mathematical formula was to be discovered. When there is the hint of a universal function in some physical law, theoretical physicists immediately look for *new physics* that will explain such a universal function. In other words, a universal function in physics is almost always connected to some sort of law and physicists of the time knew very well that some new law was needed to explain the black body radiation spectrum.

One of Kirchhoff's student, Max Planck[5], was one of these theoretical physicists engaged in the search for this new law. He was the archetype of the classical physicist and was very resistant to new ideas: he resisted, for example, the proposal that matter was composed of atoms and therefore resisted all the results connected to the statistical studies of the collection of particles pioneered by Boltzmann and Maxwell. We will shortly see that, notwithstanding his resistance, Planck had to use new concepts outside the realm of classical physics to give a solid foundation to his explanation of the black body radiation spectrum.

As we have already mentioned, we will try to be as close as possible to the history of the discovery of the distribution function for the energy density of a black body (or cavity) radiation.

Let's first see what Planck assumed in his calculations. Planck assumed that the black body is a cavity of reflecting walls with e.m. waves inside the cavity in thermal equilibrium with the walls. The thermal equilibrium can be assumed to be reached by means of small particles inside the volume thermalizing the radiation. With this assumptions, a blackbody with absorbing walls is equivalent to a blackbody that has reflecting walls and containing particles used to thermalize the radiation. He needed a more detailed physical model: the walls are composed of charged oscillators which respond to the e.m. waves inside the cavity. This is the simplest assumption that can be made, and it is justified by the assumption of thermal equilibrium between

---

[4]Gustav Robert Kirchhoff (12 March 1824 to 17 October 1887) is best known for his laws in electrical circuits and spectroscopy.

[5]Max Karl Ernst Ludwig Planck (23 April 1858 to 4 October 1947) is the founder of quantum mechanics and won the Nobel prize in 1918.

radiation and matter. There is no need to consider more complicated models of the matter composing the walls.

The next step consists in studying how an oscillating charge, for example of charge e and mass m, emits e.m. radiation when set in motion. Classical e.m. theory predicts that a charge emits radiation with power given by the Larmor formula:

$$P = \frac{e^2 \ddot{r}^2}{6\pi\varepsilon_0 c^3} \tag{3.1}$$

Eq. 3.1 is in SI units and is valid for velocities that are small compared with the speed of light[6]. Such an assumption can be assumed to be valid for the oscillators in the walls of the blackbody.

Having established that accelerated charges emit e.m. radiation, we infer that, in turn, e.m. radiation accelerates charges when interacting with them. It is then natural to assume that, when an accelerated charge emits radiation, the radiation just emitted has some influence to the charge itself. Therefore, in order to properly describe the e.m. radiation interacting with oscillators we need to calculate such a back-reaction influence. We can model this influence with an effective reaction radiation force F_{rad} that the radiation emitted by an accelerated charged particle exerts on itself (radiation reaction). Assuming a periodic motion of the oscillator with period T, from classical mechanics we know how to calculate the force associated with power P:

$$\int_0^T F_{rad} \cdot v \, dt = -\int_0^T P \, dt \tag{3.2}$$

where $v = v(t)$ is the velocity of the charge. Inserting in eq. 3.2 the Larmor power 3.1 we have:

$$\int_0^T F_{rad} \cdot v \, dt = -\int_0^T \frac{e^2 \ddot{r}^2}{6\pi\varepsilon_0 c^3} dt = -\int_0^T \frac{e^2}{6\pi\varepsilon_0 c^3} \frac{dv}{dt} \cdot \frac{dv}{dt} dt = \int_0^T \frac{e^2 \dddot{r}}{6\pi\varepsilon_0 c^3} \cdot v \, dt \tag{3.3}$$

where the last integral is evaluated by parts. The last integral contains the radiation reaction force, also called Abraham-Lorentz force, given by:

$$F_{rad} = \frac{e^2 \dddot{r}}{6\pi\varepsilon_0 c^3} = \frac{e^2 \dot{a}}{6\pi\varepsilon_0 c^3} \tag{3.4}$$

where a is the acceleration. Eq. 3.4 is a rare example in physics of third time derivative of the position or time derivative of acceleration. We are now in a position to write the equation governing the dynamics of an oscillating charge in the presence of

---

[6]Larmor formula can be made relativistically invariant to give $P = \frac{e^2 \gamma^5}{6\pi\varepsilon_0 c^3} \left(\dot{r}^2 - \frac{(\dot{r} \times \ddot{r})^2}{c^2} \right)$ where $\gamma = 1/\sqrt{1 - \frac{\dot{r}^2}{c^2}}$. It can be shown that e.m. radiation by accelerating charges is a consequence of the existance of a cosmic speed limit c.

radiation reaction. Since we are dealing with an oscillator responding to an incident e.m. radiation of the form $F = eE_0 \cos \omega t$, the equation of motion is:

$$\ddot{r} + \omega_0 r - \frac{e^2}{6\pi\varepsilon_0 c^3}\dddot{r} = \frac{eE_0}{m}\cos \omega t \tag{3.5}$$

where ω_0 is the natural oscillation frequency when there is no external force applied. We can simplify eq. 3.5 further by noticing that if the radiation reaction force is small: there is a c^3 factor at the denominator. In addition, we can simplify the algebra by working in Gauss units where the electric charge e can be written as $q = \frac{e}{\sqrt{4\pi\varepsilon_0}}$. We can therefore write:

$$\frac{e^2}{6\pi\varepsilon_0 c^3}\dddot{r} = \frac{2}{3}\frac{q^2}{c^3}\dddot{r} \simeq \frac{2q^2}{3c^3}\omega^2 \dot{r} = \gamma\dot{r} \tag{3.6}$$

and we can finally write eq. 3.5 in a more familiar form:

$$\ddot{r} - \gamma\dot{r} + \omega_0^2 r = \frac{eE_0}{m}\cos \omega t \tag{3.7}$$

This equation represents an oscillator subject to a driving force $\frac{F}{m} = \frac{eE_0}{m}\cos \omega t$ in a viscous medium exerting a force proportional to the velocity $\dot{r}$. The solution of eq. 3.7 is described by an oscillatory function whose amplitude is:

$$A(\omega) = \frac{eE_0}{\sqrt{m^2(\omega_0^2 - \omega^2)^2 + (\gamma\omega)^2}} \tag{3.8}$$

The oscillator has an average energy of $\langle E \rangle = \frac{1}{2}m\omega_0^2 A^2$ (not to be confused with E_0 the amplitude of the electric field). If the γ term is small (see fig. 3.1), i.e. small damping, the function A has a sharp maximum around the natural frequency ω_0.

Under these assumptions, we can simplify the expression of energy for $\omega \to \omega_0$:

$$\begin{aligned}
\langle E \rangle &= \frac{1}{2}m\omega_0^2 A^2 \\
&= \frac{1}{2}m\omega_0^2 \frac{e^2 E_0^2}{m^2(\omega_0^2 - \omega^2)^2 + (\gamma\omega)^2} \\
&= \frac{1}{2}m\omega_0^2 \frac{e^2 E_0^2}{m^2[(\omega_0 + \omega)(\omega_0 - \omega)]^2 + (\gamma\omega)^2} \\
&\approx \frac{1}{2}m\omega_0^2 \frac{e^2 E_0^2}{4m^2\omega_0^2(\omega_0 - \omega)^2 + (\gamma\omega_0)^2} \\
&= \frac{1}{8m}\frac{e^2 E_0^2}{(\omega_0 - \omega)^2 + (\frac{\gamma}{2m})^2}
\end{aligned} \tag{3.9}$$

where we have used the fact that, when $\omega \to \omega_0$ we can approximate $(\omega_0^2 - \omega^2) = (\omega_0 + \omega)(\omega_0 - \omega) \approx 2\omega_0(\omega_0 - \omega)$.

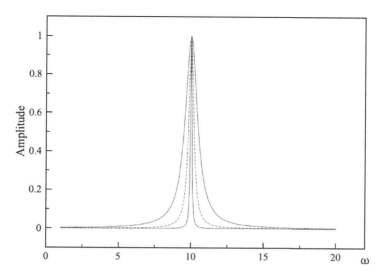

Figure 3.1 The amplitude $A \propto \frac{1}{\omega^2 + a^2}$ of a damped harmonic oscillator versus the damping factor a. As the value of the damping factor is reduced (in the figure, we have $a = 0.5, 0.2, 0.05$), the amplitude shows a sharper maximum around the resonant frequency $\omega_0 = 10$.

If we now want to calculate the average energy for all the modes in the cavity, we need to integrate eq. 3.9. When the damping factor is small (see fig. 3.1), the only values contributing to the integral are those close to the center frequency ω_0. Using a well-known integral:

$$\int_0^{\infty} \frac{d\omega}{\omega^2 + a^2} = \frac{\pi}{2a} \tag{3.10}$$

after a bit of algebra, and considering the 3 components of the field in the cavity – which gets rid of a factor 3 in the denominator – we finally obtain:

$$u(v, T) = \frac{8\pi v^2}{c^3} \langle E(v, T) \rangle \tag{3.11}$$

The relationship 3.11 is quite remarkable: it does not contain the mass m or the charge e of the oscillating charged dipole that we started with. The only information left about the original oscillator is its average energy $\langle E(v, T) \rangle$. This is a consequence of the extraordinary power of classical thermodynamics.

Notice that a straightforward application of the equipartition theorem in eq. 3.12, i.e. $\langle E(v, T) \rangle = kT$, reproduces exactly the Rayleigh-Jeans equation 2.69 of the previous chapter. Notice also that Planck could have obtained eq. 2.69 5 years earlier. In fact, Rayleigh-Jeans published this result in 1905.

Since thermodynamics brought us so far, we can try and go a bit further. Can we use thermodynamics to express the average energy? For a system with constant volume, like our cavity, we have:

$$dU = T dS \tag{3.12}$$

The above relation is valid for a thermodynamic systems but can be equally applied to a single harmonic oscillator because entropy S and internal energy U are extensive thermodynamic variables[7]. If Planck succeeded in finding an expression form the entropy of the oscillator, then the energy can be found by inverting eq. 3.12:

$$\left(\frac{\partial S}{\partial U}\right)_V = \frac{dS}{d\langle E \rangle} = \frac{1}{T} \tag{3.13}$$

where we can identify U with the average energy $\langle E \rangle$ of the oscillator.

Planck wanted to continue the road of classical thermodynamics and proposed an expression for the entropy of the oscillator:

$$S = -\frac{\langle E \rangle}{\beta v} \ln \frac{\langle E \rangle}{a v e} \tag{3.14}$$

where e is the base of natural logarithms (and not the electron charge!), v is the frequency and a and β are constants. Planck wrote eq. 3.14 in a paper written in the year 1900. He assumed this functional form for the entropy based only on the attempt at describing the blackbody radiation inside a cavity by Wien. Wien's empirical formula 2.70 is reported here for clarity as a function of the frequency v::

$$f(\frac{v}{T}) = \alpha v^3 e^{-\frac{\beta v}{T}} \tag{3.15}$$

Notice that the Rayleigh-Jeans law appeared only in 1905 and so Planck did not have a theoretical foundation for the low-frequency shape of the blackbody spectrum. Planck arrived at the definition of entropy 3.14 by equating eqs. 3.15 and 3.11. The entropy 3.14 is obtained after a bit of algebra and writing $a = \frac{\alpha c^3}{8\pi}$.

The entropy definition 3.13 reflects the fact that Wien's law 3.15 must be consistent with the second law of thermodynamics. In fact, the second derivative:

$$\frac{\partial^2 S}{\partial U^2} = -\frac{1}{\beta v}\frac{1}{U} = -\frac{1}{\beta v}\frac{1}{\langle E \rangle} \tag{3.16}$$

is always negative, β, v and $\langle E \rangle$ being always positive. At the same time as Planck was writing these expressions for the entropy of an oscillator in a cavity in thermal equilibrium with radiation, two experimental physicists, Rubens and Kurlbaum[8] (see fig. 3.2) [36], made accurate measurements of the power emitted by blackbody in the long wavelength region of the spectrum showing clearly that Wien's law was

---

[7]Extensive variables are additive for subsystems. Examples are Volume, Mass, Entropy and Energy.

[8]Heinrich Rubens (1865-1922) and Ferdinand Kurlbaum (1857-1927) were two German physicists best known for their accurate measurements of blackbody power spectrum.

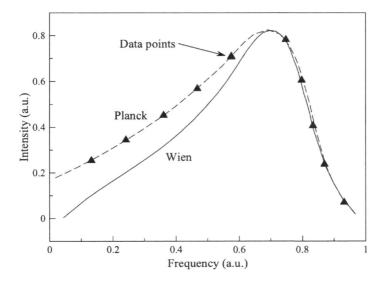

Figure 3.2 Rubens-Kurlbaum experimental data points versus Planck (dashed line) and Wien (continuous line) distribution functions.

inadequate. It seems that Rubens and his wife were having dinner at Planck's home one day. During the dinner, Rubens told Planck about his very recent measurements of the power emitted by a blackbody at long wavelengths with respect to the peak and in this occasion he reported the news to Planck exactly when he was working on his theoretical formula. Rubens gave Planck a very important information: according to his measurements, when $\frac{v}{T} \to 0$, the measurements were consistent with $\langle E \rangle \propto T$ while Wien's formula clearly showed that, for $\frac{v}{T} \to \infty$, $\langle E \rangle \propto v^3$ independent of the temperature T.

In view of this new information, Planck realized that he had to modify his entropy formula to make sure that it is consistent with the two limiting regimes when $\frac{v}{T} \to \infty$ and when $\frac{v}{T} \to 0$. From eq. 3.11, eq. 3.13 and the information from Rubens that $E \propto T$. In the regime when $\frac{v}{T} \to 0$ we must have:

$$\frac{dS}{d\langle E \rangle} \propto \frac{1}{\langle E \rangle}$$
$$\frac{d^2 S}{d\langle E \rangle^2} \propto \frac{1}{\langle E \rangle^2}$$

(3.17)

In the other regime $\frac{v}{T} \to \infty$, we have seen from eq. 3.16 that

$$\frac{d^2 S}{d\langle E \rangle^2} \propto \frac{1}{\langle E \rangle}$$

(3.18)

It is easy to see that the simplest expression for the second derivative $\frac{d^2S}{d\langle E\rangle^2}$ satisfying the two limits 3.17 and 3.18 is:

$$\frac{d^2S}{d\langle E\rangle^2} = -\frac{a}{\langle E\rangle(b+\langle E\rangle)} \tag{3.19}$$

where a and b are two constants that need to be determined from the experimental data. In fact, when $\langle E\rangle \gg b$, the constant b can be neglected and the second derivative is proportional to $1/\langle E\rangle^2$. On the other hand, when $\langle E\rangle \ll b$, the term $\langle E\rangle$ can be neglected in the parenthesis and the second derivative is proportional to $1/\langle E\rangle$.

It is now possible to integrate once eq. 3.19:

$$\frac{dS}{d\langle E\rangle} = -\int \frac{a\,d\langle E\rangle}{\langle E\rangle(b+\langle E\rangle)} \tag{3.20}$$

after the integration, and using eq. 3.13, we have:

$$\frac{1}{T} = -\frac{a}{b}\ln\left(\frac{\langle E\rangle}{b+\langle E\rangle}\right) \tag{3.21}$$

We can now extract $\langle E\rangle$ from eq. 3.21 to obtain:

$$\langle E\rangle = \frac{b}{e^{\frac{b}{aT}} - 1} \tag{3.22}$$

Having determined $\langle E\rangle$, we can now use eq. 3.11 to find the energy density in the blackbody cavity:

$$u(\nu,T) = \frac{8\pi\nu^2}{c^3}\langle E\rangle = \frac{8\pi\nu^2}{c^3}\frac{b}{e^{\frac{b}{aT}} - 1} \tag{3.23}$$

This is Planck's law. In order to have consistency with Wien's law, the two constants a and b are such that Wien's law is written as:

$$u_{Wien}(\nu,T) = \frac{8\pi\nu^3}{c^3}be^{-\frac{a\nu}{T}} \tag{3.24}$$

Notice that consistency between Planck's and Wien's laws require also that $b \propto \nu$ and has the physical dimensions of energy. Planck was able to obtain the entropy by integrating eq. 3.20. After a little bit of algebra he was able to write:

$$S = -a\left[\frac{E}{b}\ln\frac{E}{b} - \left(1 + \frac{E}{b}\right)\ln\left(1 + \frac{E}{b}\right)\right] \tag{3.25}$$

The entropy 3.25 is consistent with Planck's law which was demonstrated to be in excellent agreement with the data (see fig. 3.2). Planck now needed a theoretical justification for eq. 3.25. Being a classical physicist he tried to justify it using classical thermodynamics. However, as we now know, no matter how much effort he spent, there is no way to explain the blackbody spectrum with classical mechanics. Planck then attacked the problem by investigating the relationship between entropy and probability.

3.2 ENTROPY AND PROBABILITY

The basic problem consisted in studying how energy is distributed among a set of electromagnetic oscillator inside a blackbody cavity. Being the number of oscillators very big, it is necessary to use statistical methods. Ironically, Planck was an adversary of Ludwig Boltzmann's statistical mechanics ideas not even believing in the existence of atoms. But in this case, and in an act of desperation, he used Boltzmann's ideas.

In 1877 Boltzmann proposed[9] his famous relationship between entropy and probability:

$$S = k \ln W \tag{3.26}$$

where $k = 1.38 \times 10^{-23}$ J/K is Boltzmann's constant[10] and W is the probability of the corresponding state, i.e. the number of *distinct* microscopic states that are available to the thermodynamic system. We have already seen that the average thermal energy associated to a microscopic internal degree of freedom of a particle in a thermodynamic system is $\frac{1}{2}kT$.

Let's call ε an energy "element" associated with one of the oscillators in the blackbody cavity. We want to study how the total energy E_{tot} is distributed among N oscillators, where we expect N to be a large number. Suppose we have a number P of such energy elements ε and we want to study how they distribute among the N oscillators.

This is a direct statistical question equivalent to ask the following: how many different ways we can distribute P (identical) objects into N boxes? It turns out, as explained by Paul Ehrenfest and Heike Kamerlingh Onnes [13], that this number is:

$$W = \frac{(P+N-1)!}{P!(N-1)!} \tag{3.27}$$

If $P, N \gg 1$ eq. 3.27 becomes:

$$W \simeq \frac{(P+N)!}{P!N!} \simeq \frac{(N+P)^{(N+P)}}{P^P N^N} \tag{3.28}$$

using eq. 3.26, the entropy is:

$$S = k \ln W = k \left[(N+P)\ln(N+P) - P\ln P - N\ln N \right] \tag{3.29}$$

but we know that $P = \frac{N\langle E \rangle}{\varepsilon}$ and we have, for a single oscillator, that its entropy is:

$$\frac{S}{N} = k \left[\left(1 + \frac{\langle E \rangle}{\varepsilon} \right) \ln \left(1 + \frac{\langle E \rangle}{\varepsilon} \right) - \frac{\langle E \rangle}{\varepsilon} \ln \frac{\langle E \rangle}{\varepsilon} \right] \tag{3.30}$$

---

[9]In reality, it was Planck who wrote first the entropy in this form which can be seen on Boltzmann's grave in Zentralfriedhof in Vienna.

[10]Boltzmann's constant physical (J/K) units suggest that it relates energy and temperature. The perfect gas law $pV = nRT$ relates the pressure and volume of n moles of a perfect gas at temperature T through the gas constant R. This law can be rewritten using Boltzmann's constant as $pV = NkT$ where N is the total number of particles in one mole, i.e. the Avogadro number 6.022×10^{23}.

We can now compare eq. 3.30 with eq.3.25. The two equations are the same, providing that $b = \varepsilon$. But we also know that $b \propto v$ and therefore we must have:

$$\varepsilon = hv \tag{3.31}$$

where $h = 6.626 \times 10^{-34}$ J/Hz, is the Planck's constant. With the help of eq. 3.31 Planck was finally able to write the formula for the spectral emission of a blackbody that contained his constant h:

$$B(v, T) = \frac{2hv^3}{c^2} \frac{1}{e^{\frac{hv}{kT}} - 1} \quad \left[\frac{W}{m^2 \text{ sr Hz}} \right] \tag{3.32}$$

or, in terms of wavelengths:

$$B(\lambda, T) = \frac{2hc^2}{\lambda^5} \frac{1}{e^{\frac{hc}{k\lambda T}} - 1} \quad \left[\frac{W}{m^3 \text{ sr}} \right]. \tag{3.33}$$

At this point, we might be tempted to let $\varepsilon \to 0$, i.e. $h \to 0$, and stay with classical physics. We can be certain that Planck wanted this to be possible, being himself the last of the great classical physicists. This unfortunately will lead to a divergent entropy and so agreement is guaranteed if and only if $\varepsilon = hv$ is finite. Planck interpreted this amount of energy as a "quantum" of energy: the e.m. radiation is exchanged with oscillators in the cavity only in finite amount, thus the term "quantum". In addition, this minimum amount is proportional to the frequency of the e.m. radiation inside the cavity.

As a sanity check, let's verify that Planck formulas 3.32 and 3.33 are consistent with the two approximations of Wien and Rayleigh-Jeans. In fact, in the high-frequency regime, we have that $hv \gg kT$:

$$B(v, T) = \frac{2hv^3}{c^2} \frac{1}{e^{\frac{hv}{kT}} - 1} \simeq \frac{2hv^3}{c^2} e^{-\frac{hv}{kT}} \tag{3.34}$$

which is the Wien's formula for high frequencies (short wavelengths). If $hv \ll kT$ the exponential $e^{\frac{hv}{kT}} \simeq 1 + \frac{hv}{kT}$ and it follows that:

$$B(v, T) = \frac{2hv^3}{c^2} \frac{1}{e^{\frac{hv}{kT}} - 1} \simeq \frac{2v^2}{c^2} kT \tag{3.35}$$

which is the Rayleigh-Jeans formula for low frequencies (long wavelengths). Notice that in eqs. 3.34 and 3.35, the missing π means that they are spectral irradiance and not spectral emittance as shown in eq. 2.4. We need to check that, if $h \to 0$ we recover classical physics. In fact, Rayleigh-Jeans approximation 3.35 does not contain h as it was obtained with classical physics assumptions. Notice also that Wien's approximation does contain h since it was obtained empirically and so it should not be regarded as based on classical physics.

Planck interpreted eq. 3.31 as follows: in a blackbody cavity the radiation is in thermal equilibrium with the elementary oscillators; energy is exchanged in "packets" – or quanta – of hv and not continuously as a classical physicistwould expect.

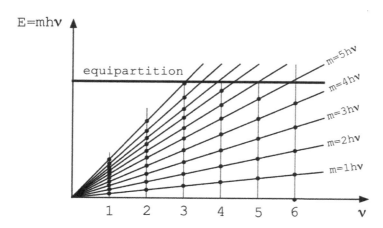

Figure 3.3 How energy is distributed among various e.m. modes inside a cavity. Classical equipartition theorem assign a constant value to all frequencies leading to a divergent total energy. Planck's formula $E = h\nu$ shows that the number of modes – shown by the black dots – decreases with increasing the frequency thus keeping the total energy constant.

Planck constant h, being very small but not zero, gives the scale of quantum phenomena.

Having the energy between the e.m. waves and the oscillators in the cavity being exchanged in discrete quanta of energy $E = mh\nu$, where m is an integer number, as the frequency increases less and less vibration modes are excited (black dots in fig. 3.3) and the ultraviolet catastrophe does not occur.

3.3 ALTERNATIVE DERIVATIONS OF PLANCK BLACKBODY FORMULA

Having followed the long, classical, route to Planck's blackbody radiation spectrum formula, we now show how it is possible to derive the formula with more direct methods. Let's go back to eq. 2.45 giving the squared magnitude of the wave vector. Let's rewrite eq. 2.45 as:

$$|k| = \pi \sqrt{\left(\frac{n_x}{L}\right)^2 + \left(\frac{n_y}{L}\right)^2 + \left(\frac{n_z}{L}\right)^2} \tag{3.36}$$

In a similar line of reasoning as in chapter 2, eq. 3.36 describe a sphere of radius $\frac{|k|}{\pi}$ in the n-space where n_x, n_y and n_z are positive integers. The volume accessible is therefore the octant where $(n_x, n_y, n_z) > 0$. The volume containing these modes is therefore:

$$V = \frac{1}{8} \frac{4\pi(\frac{|k|}{\pi})^3}{3} = \frac{4\pi\nu^3}{3c^3} \tag{3.37}$$

where we used the definition $k = \frac{2\pi v}{c}$. We now introduce the concept of *density of states*, i.e. the number of modes per unit volume. In general, the density of states is defined as:

$$g(v) = \frac{dV}{dv} \tag{3.38}$$

since we are dealing with e.m. waves, we know that there are 2 polarizations per modes and so the e.m. wave density of states is:

$$g_{em}(v) = 2\frac{dV}{dv} = \frac{8\pi v^2}{c^3} \tag{3.39}$$

Let's express the energy associated to the modes by rewriting eq. 3.36 in terms of frequency:

$$v = \frac{c}{2}\sqrt{\left(\frac{n_x}{L}\right)^2 + \left(\frac{n_y}{L}\right)^2 + \left(\frac{n_z}{L}\right)^2} \tag{3.40}$$

Using Planck formula $E = mhv$, we have that the energy associated with the modes is:

$$E = \frac{mhc}{2}\sqrt{\left(\frac{n_x}{L}\right)^2 + \left(\frac{n_y}{L}\right)^2 + \left(\frac{n_z}{L}\right)^2} \tag{3.41}$$

Eq. 3.41 tells us that we can treat each mode as a "particle" (or quantum) of energy $\varepsilon = hv$ and apply statistical physics. The index m tells us how many quanta are present in the cavity at a certain frequency v. We also consider the particles as distinguishable.

Having made the previous assumptions, we now need to calculate what is the probability that a mode with frequency v and m quanta has a certain energy E. Following McDowell[27], we treat the simplest case of a closed system[11] containing a large number of distinguishable particles N in thermal equilibrium with its surroundings at temperature T. Let us also assume that there are only two distinct energy levels (ε_0 and ε_1) each populated with, respectively, n_0 and n_1 particles, where $N = n_0 + n_1$. We now ask: how many different ways can we realize this particular configuration? Let's call W this number. Simple probability tells us that:

$$W = \frac{N!}{n_0!n_1!} \tag{3.42}$$

using Boltzmann entropy formula, we have that the entropy is:

$$S = k[\ln(N!) - \ln(n_0!) - \ln(n_1!)] \tag{3.43}$$

We now transfer a small amount of energy $\delta = \varepsilon_1 - \varepsilon_0$ to the system in such a way that one particle is promoted from the lower level to the upper level. In this case we have that the lower level is depleted ($n_0 \to n_0 - 1$) and the upper level is increased

---

[11] In thermodynamics, a closed system is a system where there is no exchange of particles with the surroundings. In our example, our system has N, V and T constant.

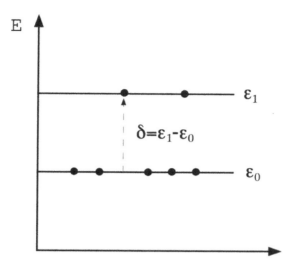

Figure 3.4 System composed by 7 distinguishable particles distributed between two energy levels ε_0 and ε_1. If the energy $\delta = \varepsilon_1 - \varepsilon_0$ is absorbed, a particle from the lower level can be promoted to the upper level,

by one unit ($n_1 \to n_1 + 1$) (see fig. 3.4). The entropy of the system after the addition of energy is:

$$S^* = k[\ln(N!) - \ln(n_0 - 1)! - \ln(n_1 + 1)!]. \tag{3.44}$$

We can calculate the entropy change of such transformation:

$$
\begin{aligned}
\Delta S &= S^* - S \\
&= k[\ln(N!) - \ln(n_0 - 1)! - \ln(n_1 + 1)! - (\ln(N!) - \ln(n_0!) - \ln(n_1!))] \\
&= k[-\ln(n_0 - 1)! + \ln(n_0)! - \ln(n_1 + 1)! + \ln(n_1)!] \\
&= k[\ln(n_0) - \ln(n_1 + 1)] \\
&= k\ln\left(\frac{n_0}{n_1 + 1}\right) \simeq k\ln\left(\frac{n_0}{n_1}\right)
\end{aligned} \tag{3.45}
$$

where the last equality assumes $n_1 \gg 1$.

In a thermodynamic process at constant volume we know that $\Delta S = \Delta Q/T$ and the change in energy ΔQ is simply the energy added δ so that $\Delta S = \delta/T$. Therefore we have:

$$\Delta S = k\ln\left(\frac{n_0}{n_1}\right) = \frac{\delta}{T} \tag{3.46}$$

from which we obtain:

$$\frac{n_1}{n_0} = e^{-\frac{\delta}{kT}} \tag{3.47}$$

If we generalize to any two arbitrary energy levels separated by the energy E, we obtain the *Boltzmann distribution law*:

$$\frac{n_1}{n_0} = e^{-\frac{E}{kT}} \tag{3.48}$$

relating the number of particles n_1 in an energy level E_1 separated by an energy E from a lower energy level E_0 with n_0 particles and with $E = E_1 - E_0$. If we normalize the Boltzmann function such that:

$$\int_0^\infty Ae^{-\frac{E}{kT}} = 1 \tag{3.49}$$

the function $f(E) = Ae^{-\frac{E}{kT}}$ can be interpreted as the probability that a particle has energy E, where the energy E is a continuous variable. The distribution function $f(E)$ is a very powerful mathematical tool. In fact, it allows us to calculate averages of physical quantities over the thermodynamic system described. The average energy $\langle E \rangle$, in the discrete case, is:

$$\langle E \rangle = \sum_0^\infty E_m \cdot f(E_m) \tag{3.50}$$

where $E_m = mh\nu$ is the discrete energy with m integer running from 0 to ∞.

We now go back to eq. 3.41 and identify the cavity modes as identical but distinguishable particles[12]. For a mode at frequency ν with associated m quanta of energy $mh\nu$, with energy E quantized, i.e. assuming only discrete values, the normalization integral 3.49 becomes a summation:

$$\sum_{m=0}^\infty Ae^{-\frac{E}{kT}} = A\sum_{m=0}^\infty (e^{-\frac{h\nu}{kT}})^m = 1 \tag{3.51}$$

Using the well-known formula for the geometric sum, for $r < 1$:

$$\sum_{n=0}^\infty ar^n = \frac{a}{1-r} \tag{3.52}$$

eq. 3.51 becomes:

$$A\sum_{m=0}^\infty (e^{-\frac{h\nu}{kT}})^m = A\left(\frac{1}{1 - e^{-\frac{h\nu}{kT}}}\right) = 1 \tag{3.53}$$

from which we see that the normalization factor A is:

$$A = 1 - e^{-\frac{h\nu}{kT}} \tag{3.54}$$

Having properly normalized the distribution function, we can now rewrite eq. 3.50 as:

$$\langle E \rangle = A\sum_{m=0}^\infty E_m \cdot f(E_m) = A\sum_{m=0}^\infty mh\nu(e^{mh\nu})^{-\frac{1}{kT}} = A\sum_{m=0}^\infty mh\nu(e^{mh\nu})^\beta \tag{3.55}$$

where $\beta = -\frac{1}{kT}$.

---

[12]This is still a classical definition.

Notice that the last summation in eq. 3.55 is the derivative of $\sum_{m=0}^{\infty}(e^{mhv})^{\beta}$ with respect to β. Therefore we can write:

$$\langle E \rangle = A \frac{d}{d\beta} \sum_{m=0}^{\infty} (e^{mhv})^{\beta} = A \frac{d}{d\beta} \left(\frac{1}{1 - e^{-\beta hv}} \right) \tag{3.56}$$

Evaluating the derivative and inserting the value for A from eq. 3.54, we have:

$$
\begin{aligned}
\langle E \rangle &= A \frac{d}{d\beta} \left(\frac{1}{1 - e^{-\beta hv}} \right) \\
&= -\left(1 - e^{-\beta hv}\right) \left(\frac{1}{1 - e^{-\beta hv}} \right)^2 \cdot -hv \cdot e^{-\beta hv} \\
&= \frac{hv}{e^{\frac{hv}{kT}} - 1}
\end{aligned}
\tag{3.57}
$$

and we obtain Planck formula by inserting eq. 3.57 into eq. 3.11.

We conclude this section by pointing out that the other Planck formula $E = hv$ is key for the transition from classical to quantum physics. So, the energy is exchanged between the e.m. field and the oscillators not continuously but in chunks, called quanta. The attentive reader might already have noticed that in a few places in the previous analysis, the equations of statistical physics – normally used for particles – have been used for the mode of vibrations of the e.m. radiation inside the cavity.

3.4   HERTZ DISCOVERY OF E.M. WAVES

We have seen in chapter 2 that the laws of magnetism and electricity were unified by Maxwell's equation 2.1 around the year 1865. These equations were explaining more or less all the e.m. phenomena known at the time. However, as it is the case with good theories, the equations also predicted the existence of e.m. waves propagating at a constant speed $c = 299,792,458$ meters/sec. This prediction started a race among experimentalists to generate and detect such e.m. waves. Among the many working on it, the first to be successful was the German physicist Heinrich Hertz in 1886[13].

The apparatus used by Hertz is shown schematically in fig. 3.5. It consists of two parts: on the left there is the generating and transmitting part; on the right the receiving part. In between the two parts, a shield made of different material can be inserted to study the propagation through it. The general idea consisted in generating sparks between the spark gaps of the transmitting dipole antenna. The dipole antenna is made of two brass spheres (acting as capacitors) at the end of a conducting wire. The wire is interrupted by two small conducting spheres separated by a small gap. If the transmitting antenna is part of a resonant circuit, the spark will be able to

---

[13]Heinrich Rudolph Hertz (1857-1894) was a German physicist who first proved the existence of e.m. waves. The SI unit of frequency – the Hertz abbreviated Hz – was established in his honor.

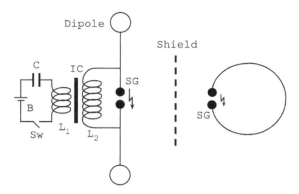

Figure 3.5 Simplified schematics of the electrical circuit used by Heinrich Hertz to generate and detect e.m. waves. A resonant circuit consisting of a battery B, a capacitor C and an induction coil – made of two coupled inductors L_1 and L_2 and an iron core IC – is connected to a dipole antenna containing a spark gap. Sparks generated are received by a loop antenna containing another spark gap.

excite the resonant frequency thus generating e.m. waves at that specific frequency. The resonant circuit of the left part of the circuit is designed in such a way that the voltage generated on the L_2 output of the induction coil is high enough to exceed the breakdown voltage of the air in between the spark gap[14]. Every time the switch is operated, a spark is generated and a burst of damped e.m. waves is produced. The waves then propagate toward the receiving loop where, if properly tuned, will generate smaller sparks between the receiving spark gaps. Hertz was able to see sparks received even if the transmitting antenna was several meters away.

Hertz was able to measure various properties of the e.m. waves, including their ability to reflect, refract, their polarization, etc. In an effort to see better the tiny sparks generated at the receiving antenna, Hertz enclosed the receiving antenna in a box made of some material transparent to e.m. waves, like, for example, cardboard. When he did this, he found that the sparks were reduced greatly thus suggesting that the light was somehow "helping" the generation of sparks. In a series of experiments, and by using a prism to select the type of light that was incident to the receiving spark gap, Hertz determined that the light "helping" the generation of spark was in the ultraviolet region. Ultraviolet light, but not light at lower wavelengths, was able to extract electrons from the surface of the metal of the spark gaps thus increasing the length of the received sparks. Hertz did not have an explanation for this phenomenon which was later termed "photoelectric effect".

---

[14]The breakdown voltage is the minimal voltage applied to an insulator, in this case, the air in between the two small spheres of the spark gap, to generate a large reduction of resistance which, in turn, generates a sudden transfer of electrons in the form of a spark.

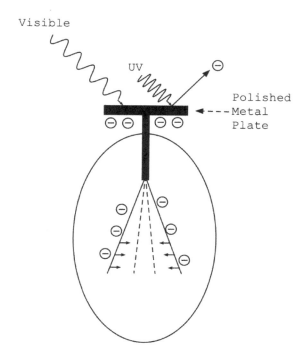

Figure 3.6 Hallwachs' experiment showing that negatively charged particles are extracted from a metal plate when ultraviolet light is shone upon it.

3.5 HALLWACHS AND LENARD EXPERIMENTS

Right after Hertz experiments showed the photoelectric effect, many researchers started to look at understanding better the phenomenon. Of particular relevance was a simple experiment by Hallwachs[15] which consisted of a gold leaf electroscope connected to a polished disk of Zinc (see fig. 3.6) through a conducting wire. When the electroscope is charged, either negatively or positively, the leaves are subject to a Coulomb electrostatic repulsion and spread apart. If connected to ground, i.e. allowing the charges to be neutralized, the leaves stop to be repelled and become limp. The electroscope is said to be discharged.

Hallwachs made a very simple experiment: he charged the electroscope first negatively and then positively while illuminating the metal plate with ultraviolet light coming from an arc lamp or burning magnesium. Hallwachs noticed a very interesting phenomenon: once charged and illuminated, the electroscope discharged only when charged negatively. When charged positively, the electroscope did not

---

[15]Wilhelm Hallwachs (1859-1922) was a German physicist that worked as an assistant of Hertz during the investigations of the nature of e.m. waves.

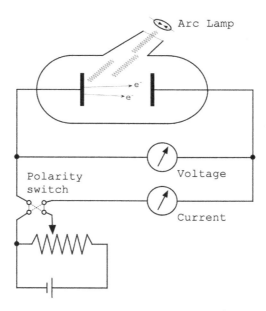

Figure 3.7 Simplified schematic of Lenard's experiment for the characterization of the photoelectric effect. A polarity switch allows the voltage applied to the two plates inside the vacuum tube to change polarity.

discharge. Illuminating the electroscope with visible light, no matter the intensity, did not discharge the electroscope.

The qualitative interpretation was very direct: ultraviolet light is capable of extracting the negative charges thus discharging the electroscope.

In 1899, J.J. Thomson[16] showed that the negative charged particles in Hertz and Hallwachs experiments were electrons. Better quantitative experiments were needed to better understand the photoelectric effect.

In 1902, one of Hertz's students, Philip Lenard[17] devised an experiment[18] to measure in detail the photoelectric effect.

In fig. 3.7 Lenard's experimental set up is shown schematically. A vacuum glass tube contains 2 separated metal electrodes facing each other. A quartz window, transparent to ultraviolet, allows light coming from a carbon arc lamp to shine directly on one of the electrodes, normally indicated as *anode*. A carbon arc lamp was used because it was known to be an intense source of ultraviolet light.

---

[16] Sir Joseph John Thomson (1856-1940) was a British physicist who discovered the electron by showing that the cathod rays were small particles, negatively charged and with a large charge to mass ratio.

[17] Philip Eduard Anton Von Lenard (1862-1947) was a German physicist who won the Nobel Prize for physic in 1905 for his work on cathode rays.

[18] See [44] for a detailed historical recount of Lenard's work on photoelectric effect.

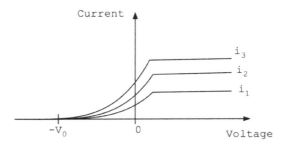

Figure 3.8 Voltage/current characteristics of the photoelectric current for three different light intensities $i_1 < i_2 < i_3$. Notice that the stopping potential $-V_0$ does not change with changing the light intensity.

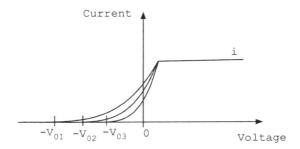

Figure 3.9 Voltage/current characteristics of the photoelectric current for three different frequencies of the light .

Lenard was able to control 1) the intensity of the light coming out of the arc lamp, 2) the potential difference ΔV applied between the cathode and the other electrode (anode) and 3) the polarity of this potential difference. When there was no potential difference applied between electrodes ($V = 0$ in fig. 3.8), a current flowing in the circuit is already observed, and it is increasing with increasing the intensity of the light coming from the arc. Keeping fixed the light from the arc lamp, an increase in the current (also called *photo current*) with increasing the potential difference, is observed until saturation, i.e. the condition for which increasing the potential difference does not increase the photo current. In fig. 3.8 the three curves shown correspond to three different intensity of light saturating at the different currents labeled i_1, i_2 and i_3. Inverting the polarity of the potential difference had the effect of decreasing the photo current to zero, no matter what is the intensity of the light from the arc lamp. This specific potential difference $(-V_0)$ is called *stopping potential*.

Lenard had also enough intensity in the light to the point that he was able to pass it through a prism. He was able to verify that the photocurrent was generated only if the frequency of light was in the ultraviolet region of the spectrum, as already determined by Hertz. He went a bit further and found that the stopping potential depended on the frequency of the light and not the intensity, as shown in fig. 3.9. In addition, he also found that the saturation current depends only on the intensity of light but not the frequency, i.e. given a certain light intensity the saturation current is the same.

Classical physics was not able to describe all these experimental data. In particular, there are three different observations that cannot be explained and they are: 1) no lag time between the absorption of e.m. radiation by the anode and the emission of the photoelectron; 2) the kinetic energy of the emitted photoelectron does not depend on the intensity of the light and 3) photoelectrons are not emitted if the frequency of the light is less than a cut-off frequency, no matter how intense is the light shining on the anode.

We can attempt at making a classical model of the photoelectric effect. Let's assume classically that electrons are bound inside the metal and a minimum amount of energy E_0 is needed for an electron to escape the anode. This minimum energy is referred to as *work function*. The kinetic energy of an electron, in classical terms, right after being ejected from the surface of the anode be K. This energy has been transferred from the incident e.m. wave to the electron. The electron is now free and is subject to an electric field which will increase its energy by an amount eV. If we neglect other forces like, for example, gravity then the only force acting upon the electron is the electric force. If we now apply the work-energy theorem[19] we can write:

$$K - eV = 0 \tag{3.58}$$

If we now apply the stopping potential V_0, the photoelectron will lose its initial kinetic energy K_i and it will come to a stop. The energy balance will be:

$$K_i = eV_0 \tag{3.59}$$

which means that the largest kinetic energy of the photoelectron, when a stopping potential is present, is its initial kinetic energy right after being released from the anode. Therefore, measuring the stopping potential is equivalent to measuring the largest kinetic energy of the photoelectron:

$$K_{max} = eV_0. \tag{3.60}$$

We see immediately that the above classical interpretation is in direct contradiction with the experimental data shown in fig. 3.8. In the classical view, the photoelectron absorbs the e.m. wave continuously: high intensity e.m. waves should produce correspondingly high energy photoelectrons and therefore high stopping potential. In a similar way, when the radiation intensity is low, the expected kinetic energy of the photoelectron should be low with a corresponding small stopping potential. The data show a completely different picture: the stopping potential does not depend on the intensity of the light! In fact, fig. 3.8 shows that the stopping potential $-V_0$ is the same for different light intensities ($i_3 > i_2 > i_1$). Even more puzzling is the fact that the stopping potential instead depends on the frequency of the light, as shown in fig. 3.9.

---

[19]The work-energy theorem states that the work done by the sum of all forces acting on a particle is equal to the change in the particle's kinetic energy.

3.6 EINSTEIN'S PHOTON HYPOTHESIS

In 1905 Albert Einstein, a young physicist working in the Swiss patent office in Bern, published in the journal Annalen der Physik a revolutionary paper titled "Concerning an heuristic point of view toward the emission and transformation of light[20].

We have seen previously that Planck was able to explain the spectral distribution of the blackbody radiation by assuming that the elementary charged oscillators emit and absorb e.m. radiation in quanta of energy $E = h\nu$. Einstein's went a step forward and assumed it is the e.m. radiation itself that is quantized in units called *photons*. In his paper Einstein, using thermodynamic and statistical considerations, makes the hypothesis that e.m. radiation is in fact consisting of a number of discrete corpuscles of energy $h\nu$. This photon hypothesis naturally explained all the measurements described in the previous paragraph. According to Einstein, the e.m. UV radiation in the photoelectric effect penetrates into the surface layer of the metal and is totally absorbed and transformed in kinetic energy of the electrons. Each photon therefore delivers its total energy to a single electron, i.e. one single photon delivers its energy to one single electron. An electron that is embedded in the interior of the metal electrode will need to lose some of its initial energy before it can reach the surface. The electrons residing on the metal surface will be emitted with the largest velocity with kinetic energy:

$$E = h\nu - P \tag{3.61}$$

where P is the amount of work that the electron must perform to leave the metal surface, according to Einstein. Eq. 3.61 explains why in Lenard's experiments the energy of the photoelectrons emitted show no dependence on the intensity of the light. In fact, the basic assumption that one photon ejects one electron means that the frequency of the photons and not the their number, i.e. the intensity, needs to be above a threshold $\nu_{min} = (E - P)/h$.

Eq. 3.61 predicts that the energy of each electron ejected by the metal must increase linearly with frequency of the photon with a slope equal to Planck's constant. In 1914 and 1916, R. A. Millikan[21], although convinced that Einstein's photon hypothesis was wrong because light was clearly a wave, published instead two papers where he confirmed that Einstein's photon hypothesis was indeed correct [29] [30]. He also gave a measured value of Planck's constant. Albert Einstein received the Nobel Prize in physics in 1921 for his discovery of the law of the photoelectric effect.

Can the photon be considered a particle, having its energy defined in quanta of energy $E = h\nu$? Or is it a wave because light shows all the wave phenomena like, for example, interference and diffraction? In 1923 in an important experiment by

---

[20]The original title was "Uber einen die Erzeugung und Verwandlung des Lichtes betreffenden heuristischen Gesichttspunkt". The interested non-German-speaking reader is invited to read the English translation by Arons and Pippard [3].

[21]Robert Andrews Millikan (1868-1953) was an American physicist known for being the first to measure the elementary electric charge and for verifying Einstein's predictions of photoelectric effect. He was awarded the Nobel Priz in 1923.

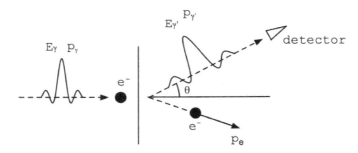

Figure 3.10 Compton scattering: an e.m. wave of energy E_γ (X-ray) and momentum p_γ, incident on an electron at rest, is scattered in a different direction, with energy $E_{\gamma'}$ (X-ray) and momentum $p_{\gamma'}$. The electron is also scattered in another direction with momentum p_e.

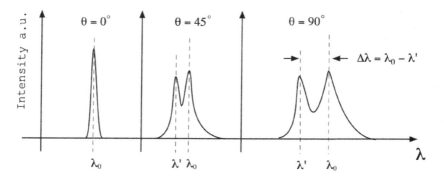

Figure 3.11 Results of the Compton scattering experiments for three different angles (0, 45 and 90 degrees). The intensity of the light is in arbitrary unit.

Arthur Compton[22] it was observed that a fraction of the X-rays scattered off a piece of graphite had longer wavelengths as shown in fig. 3.10. We can consider electrons in graphite to be free and at rest, considering the energy of the incident X-rays.

Compton observed the intensity and wavelength of the scattered photons as a function of the scattering angle θ, i.e. the angle between the incoming beam and the deflected beam. In a purely classical view, scattered light by free electrons is expected not to change its wavelength. Compton instead observed not only that the wavelength of the light scattered was different from the wavelength of the incident light, but it depended on the scattering angle (see fig. 3.11).

Compton was interested in the relationship between the wavelength shift of the scattered light versus the observing angle. He found a very interesting result: the wavelength shift $\Delta\lambda = \lambda' - \lambda_0$ was proportional to the factor $(1 - \cos\theta)$, where λ'

---

[22]Arthur Holly Compton (1892-1962) was an American physicist Nobel prize in 1927 for the so-called Compton effect described in the text.

is the wavelength of the scattered radiation and λ_0 is the wavelength of the incident radiation. Since classical physics could not explain the data, Compton used Einstein's suggestion that light comes in quanta and imagined that the scattering of X-rays was described as inelastic scattering between free electrons and photons. A beam of monochromatic light of frequency v or, equivalently wavelength λ, can be thought as composed of either a classical wave or as a beam of photons traveling at the speed of light c (in vacuum). To explain his data, Compton had to rely again to Einstein and use the relativistic energy equation of a particle:

$$E^2 = p^2c^2 + m_0^2c^4 \tag{3.62}$$

where p is the momentum of the particle and m_0 is its rest mass. In the case of the photon, its rest mass is equal to zero and so, as we had already mentioned in section 2.2, the photon has a momentum[23]:

$$p = \frac{E}{c} \tag{3.63}$$

Using Planck formula $E = hv$, eq. 3.63 becomes:

$$p = \frac{hv}{c} = \frac{h}{\lambda} \tag{3.64}$$

Next step is to assume that in the scattering we have conservation of energy (in the relativistic form 3.62) and conservation of linear momentum. With reference to fig. 3.10 the conservation of energy is:

$$E_\gamma + E_e = E_{\gamma'} + E_e' \tag{3.65}$$

where E_γ is the energy of the incoming photon, E_e is the energy of the electron at rest, $E_{\gamma'}$ is the energy of the scattered photon and E_e' is the energy of the scattered electron. Using eqs. 3.62 and 3.64, we have:

$$\frac{hc}{\lambda} + m_ec^2 = \frac{hc}{\lambda'} + \sqrt{p_e^2c^2 + m_e^2c^4} \tag{3.66}$$

where λ is the wavelength of the incoming photon, λ' is the wavelength of the scattered photon, p_e is the momentum of the scattered electron and m_e is the electron mass.

Conservation of momentum requires that:

$$p_\gamma = p_e + p_{\gamma'}. \tag{3.67}$$

The derivation continues by eliminating p_e from equations 3.66 and 3.67. This is achieved by squaring the energy 3.66 and momentum 3.67 equations. We have for

---

[23]This result is not strictly relativistic but can be deduced from Maxwell's equations.

the energy:

$$p_e^2 c^2 + m_e^2 c^4 = (\frac{hc}{\lambda} + m_e c^2 - \frac{hc}{\lambda'})^2$$

$$p_e^2 c^2 = (\frac{hc}{\lambda})^2 + (\frac{hc}{\lambda'})^2 - 2\frac{h^2 c^2}{\lambda \lambda'} + 2m_e c^2 (\frac{hc}{\lambda} - \frac{hc}{\lambda'}).$$

(3.68)

In order to square the momentum eq. 3.67, we need to consider that it is a vector equation. The modulus of the square of a vector is given by the scalar product $|v^2| = v \cdot v$. More in general, we have that given two vectors v_1 and v_2, their scalar product is $v_1 \cdot v_2 = v_1 v_2 \cos \theta$. It follows that:

$$p_e = p_\gamma - p_{\gamma'}$$

$$p_e^2 = p_e \cdot p_e = p_\gamma^2 + p_{\gamma'}^2 - 2p_\gamma p_{\gamma'} \cos \theta$$

(3.69)

Inserting eq. 3.63 into eq. 3.69 and multiplying by c^2, we have:

$$p_e^2 c^2 = \frac{h^2 c^2}{\lambda^2} + \frac{h^2 c^2}{\lambda'^2} - 2\frac{h^2 c^2}{\lambda \lambda'} \cos \theta$$

(3.70)

By finally equating eq. 3.69 with eq. 3.70 we obtain the Compton formula:

$$\Delta \lambda = \lambda' - \lambda = \frac{h}{m_e c}(1 - \cos \theta)$$

(3.71)

where the factor $\frac{h}{m_e c}$ is the *Compton wavelength* of the electron[24], i.e. the wavelength that a photon must possess such that its energy is equal to the electron mass-energy. This is easily seen by equating Planck formula $E = \frac{hc}{\lambda}$ with Einstein formula $E = mc^2$. The formula 3.71 reproduces extremely well the data. The Compton scattering formula is formally identical to the formula of inelastic scattering between solid objects like, for example, billiard balls. In this particular experiment the photon can be interpreted as being a particle which is quite revolutionary because light, including X-rays, behaves like a wave. We will see later in the book that this duality wave-particle is very important in quantum mechanics.

Having determined experimentally that the photon hypothesis is valid, we can go back to eq. 3.57 and interpret the average energy:

$$\langle E \rangle = \frac{h\nu}{e^{\frac{h\nu}{kT}} - 1}$$

(3.72)

as the average number of photons in a particular mode of frequency ν in thermal equilibrium at temperature T. We can formally rewrite eq. 3.72 as:

$$\langle E \rangle = \langle n_\nu \rangle h\nu$$

(3.73)

---

[24]The Compton wavelength for an electron is equal to approximately $2.426 \cdot 10^{-12}$ meters.

where $\langle n_V \rangle$ is the average number of photon, also called **photon occupation number**, defined as:

$$\langle n_V \rangle = \frac{1}{e^{\frac{hv}{kT}} - 1}. \tag{3.74}$$

We see that, when $\frac{hv}{kT} \gg 1$ then $\langle n_V \rangle \to e^{-\frac{hv}{kT}}$ i.e. the classical Boltzmann distribution factor 3.48 indicating that we re obtain classical physics at high frequencies and low temperatures.

3.7 SPECIFIC HEAT OF SOLIDS

Einstein's contribution to the early phases of quantum theory did not stop at the photon hypothesis. He went a bit further and tried to explain other phenomena that classical physics was not able to explain [14].

In 1809 there was experimental evidence, by Dulong[25] and Petit[26], that the specific heat capacity of some elements, when measured at ambient temperature, is a constant. The specific heat capacity, normally indicated with C[27], is the amount of heat given to one mole of the substance to increase its temperature by 1 degree therefore measured in Joule/K.

Dulong and Petit law states that the molar heat capacity *for all solids* is a constant:

$$C = \frac{\partial E}{\partial T} = 3nR \tag{3.75}$$

where $R = 8.314463$ J·K/mol is the gas constant and n is the number of moles of the substance. If instead of the number of moles we use the total number of atoms/molecules then the law of Dulong and Petit can be written as:

$$C = 3Nk \tag{3.76}$$

where N is the total number of atoms/molecules and k is the Boltzmann constant.

We can explain classically eqs. 3.75 and 3.76 by assuming that each molecule vibrates as a free harmonic oscillator with three spatial and three momentum degrees of freedom. The equipartition theorem states that each degree of freedom contributes $\frac{1}{2}kT$ to the total energy so that:

$$E = 6 \cdot \frac{1}{2}kT = 3kT \tag{3.77}$$

from which, after differentiation, we get eq. 3.76.

---

[25] Pierre Louis Dulong (1785-1838) was a French chemical physicist known for his studies in thermodynamics.

[26] Alexis Therese Petit (1791-1820) was a French physicist known for his work on the thermodynamics of steam engines.

[27] Her we intend the specific heat at constant volume.

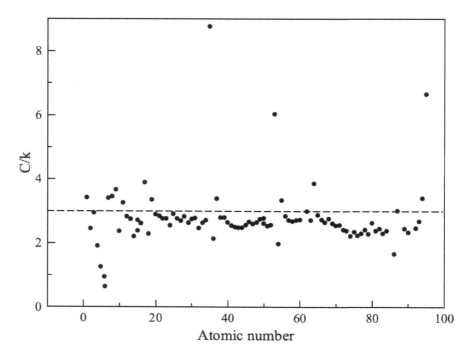

Figure 3.12   Specific heat of the elements versus atomic number at $T = 25°C$. The horizontal dashed line at $C/k = 3$ marks the Dulong-Petit law. Data from [1].

The Dulong-Petit law 3.75 or 3.76 is in reasonable agreement with experimental data, a part from some outliers, as shown in fig. 3.12, providing that the temperature is above some critical value typical of the material. When the temperature is below the critical value data show that $C \to 0$ as shown, for example, in fig. 3.13 for diamond. The linear relation 3.77, consequence of the equipartition theorem, cannot explain the failure of the Dulong-Petit law.

Einstein attempted at a theory of solid matter capable of explaining the specific heat anomaly. He tried to incorporate the new ideas of quantization of energy that successfully explained the blackbody radiation spectrum. The so-called *Einstein's model* of a solid consists of N atoms/molecules free to vibrate at a frequency $v = \frac{\omega}{2\pi}$ confined in some sort of potential well. This hypothesis comes directly from the success in explaining the blackbody radiation by considering the quantization of the exchange of energy between the oscillator in the walls of the blackbody and the radiation. As the average energy of oscillators in the blackbody cavity was not the 3kT coming from equipartition, Einstein assumed that equipartition was not to be assumed for the correct description of the specific heat of a solid.

According to Planck, eq. 3.72 gives the average energy of an oscillator. If we consider the limit where $\frac{hv}{kT} \ll 1$, i.e. low frequency and high temperature, we have

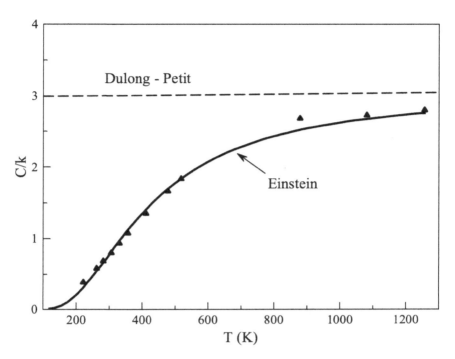

Figure 3.13 Specific heat of diamond as a function of temperature T. Data points are measurements while the line is Einstein's model.

that:

$$e^{\frac{hv}{kT}} \approx 1 + \frac{hv}{kT} + \dots \tag{3.78}$$

Using the approximation 3.78, the average energy becomes:

$$\langle E \rangle = \frac{hv}{e^{\frac{hv}{kT}} - 1} \approx kT \tag{3.79}$$

thus recovering the classical equipartition and the Dulong-Petit law (a part a factor 3 to be inserted).

If we instead use Planck's expression for the average energy, we obtain the molar heat capacity by taking the derivative:

$$C_V = \frac{\partial \langle E \rangle}{dT} = 3Nk \left(\frac{hv}{kT} \right)^2 \frac{e^{\frac{hv}{kT}}}{\left(e^{\frac{hv}{kT}} - 1 \right)^2}. \tag{3.80}$$

Eq. 3.80 predicts that the heat capacity goes to zero as the temperature goes to zero. Einstein's model explains reasonably well the data (see fig. 3.13) providing a further demonstration that quantization of energy can explain various physical phenomena.

Einstein's model, although predicting correctly the low temperature behavior of the specific heat of solids, shows deviations from experimental data due mainly to the very simplifying assumptions. All the oscillators, for example, are assumed to vibrate at the same frequency while it would be more accurate to assume a more complex spectrum of oscillations. A more accurate theory, beyond the scopes of this book, has been elaborated by Debye that uses a more accurate set of assumptions.

4 Early Quantum Theory: Bohr's Atom

In the year 1913, a Danish physicist, Niels Bohr[1], tried to extend Planck's suggestion that energy is quantized to the description of atoms. The experimental and theoretical basis for the atomic model was mainly due to the work of J. J. Thomson , E. Rutherford and N. Bohr , during the years from 1900 to 1913. Following the discovery of the electron, in 1897, in a series of experiments, Thomson discovered the electron and established a series of important properties, from the mass/charge ratio to the fact that this particle was present in all atoms and was negatively charged.

It was established that electrons can be extracted from atoms and that such action left behind a positively charged residual mass much bigger than the electron's mass. It was therefore clear that atoms are not indivisible and that electrons needed to be incorporated in the mechanics of the atom in such a way that the overall electric charge was zero. The simplest model by Thomson was the so called "plum pudding" model as depicted in fig. 4.1. In this model, the negatively charged particles (electrons) are dispersed in a cloud of positive charge so as to make the atom neutral. The electrons can vibrate around their equilibrium position and, if enough energy is provided, they can be extracted. Due to the small mass of the electron, the majority of the mass of the atom is attributed to the positive charge.

Thomson's model was not able to explain a classic experiment by E. Rutherford in 1906. Rutherford, while working in Canada in 1899, was able to identify three types of emissions from radioactive materials: alpha, beta and gamma radiation. In a series of experiments he was able to determine that the beta radiation was composed of electrons and the alpha radiation was made of Helium nuclei, much heavier than electrons and positively charged. The gamma radiation instead turned out to be highly energetic photons.

Rutherford then thought to explore the structure of matter by bombarding thin sheets of metals with alpha particles. This alpha-particles scattering work was inspired by the fact that he already observed that some alpha-particles, when directed into air, were scattered but no quantitative measurements were performed. To follow up these observations he decided to study if the scattering was present when alpha-particles were traversing solids. He took advantage of the fact that gold can be hammered down and reduced to extremely thin foils of a few tens of atoms thickness. The gold target was therefore thin enough that most of the alpha-particles were effectively traveling through the solid without significant stopping.

---

[1]Niels Bohr (1885-1962) was a Danish physicist considered one of the founding fathers of quantum mechanics. He was awarded the "Nobel Prize for physics in 1922."

DOI: 10.1201/9781003145561-4

83

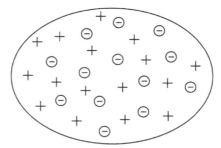

Figure 4.1 Thomson's plum pudding model of the atom. The negatively charged electrons are dispersed into a "sea" of positive charge.

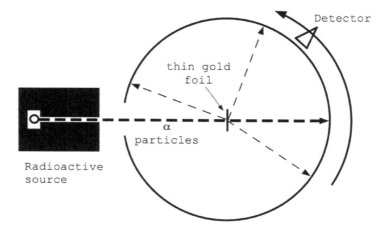

Figure 4.2 Rutherford's experiment. A collimated beam of alpha particles is directed toward a very thin gold foil. The majority of alpha particles passes through unscattered. A few alphas are scattered at various angles even at more than 90° from the original direction.

Rutherford enclosed some radioactive Radium inside a cavity made of lead and where he practiced a long thin hole (see fig.4.2). This allowed him to produce a thin collimated beam of alpha-particles. He was also aware that the alpha-particles were heavy (with respect to the electron mass) and positively charged particles. Rutherford observed that the majority of alpha-particles continued to fly along the original direction. However, and this is the puzzling observation, some particles were scattered even at large angles. It was observed that a few were even back scattered.

The results of this experiment [37], and others more sensitive performed by his collaborator Geiger, are shown in fig. 4.3. It is clear that the large scattering angle of some alpha-particles and the fact that increasing the path of the alpha particles inside the matter increases the number of large scatterers, was not accounted for the Thomson model. In Rutherford's own words [37]:

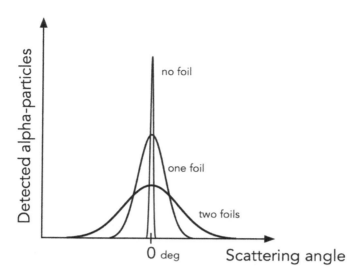

Figure 4.3 Results from Geiger's experiment showing that the scattering is increased when none, one and two gold foils are interposed on the alpha-particles beam.

"...it can easily be calculated that the change of direction of 2 degrees ... would require over that distance an average transverse electric field of about 100 million volts per cm. Such a result brings out clearly the fact that the atoms of matter must be the seat of very intense electrical forces ..."

It comes quite natural to propose, as indeed Rutherford did, a *planetary model* of the atom where the majority of the mass is concentrated in a small volume (nucleus) and the electrons orbit around. The mechanics of such an atomic model would not be different from celestial mechanics considering that the electric forces have exactly the right inverse square law for the attractive force. So, according to Rutherford's model, the majority of the mass of the atom is concentrated in the nucleus charged positively, around which negatively charged electrons orbit in circular paths. If the electric charge of the nucleus is balanced with the negative charge of the electrons then the atom is electrically neutral. The electric force is dominating with respect to the gravitational attraction and being much stronger, makes the atom relatively small.

Although the planetary model seemed to make sense, there was a fundamental problem with it: its stability. Maxwell's equations do not allow two electrical charges to orbit one around the other without emitting e.m. waves, since the charges are constantly accelerated. According to classical physics, the electron should spiral very rapidly toward the proton emitting e.m. waves in the process. Classical matter is not stable.

In addition, if atomic elements are excited, they show an e.m. emission spectrum composed of a combination of narrow lines characteristic of the element. The portion of emission spectrum of Hydrogen in the visible is shown in fig. 4.4. Considering that hydrogen is the simplest element, it is no surprise that its emission spectrum looks

Figure 4.4 Four hydrogen emission lines in the visible portion of the e.m. spectrum.

relatively simple. In fact, some researchers were able to find an empirical formula capable of predicting the wavelength of the hydrogen lines with excellent agreement with the data. In 1880 J. Rydberg found that his proposed empirical formula was able to predict well the emission spectrum of most alkali metals:

$$\frac{1}{\lambda} = R_H \left(\frac{1}{n_1^2} - \frac{1}{n_2^2} \right) \tag{4.1}$$

where $R_H = 1.09678 \cdot 10^{-2}$ nm^{-1} is the Rydberg constant and n_1 and n_2 are two integers. Inherent in the formula 4.1 is the Rydberg-Ritz *combination principle* according to which the spectral lines of any element include either the sum or the difference of other two lines in the same spectra.

Niels Bohr, having recently learned about Planck quantization of the way oscillators exchange energy with e.m. waves, made an attempt at proposing an atomic model incorporating the new ideas on the quantization of e.m. radiation. He was not satisfied with the planetary model of the atom for many reasons. In his Nobel lecture[5] he made clear that there are serious difficulties with the planetary model and the measured properties of atoms. The first and certainly important difference consists in the fact that in a gravitational bound system the motion of the bodies is not completely determined by the law of gravitation but strongly depends on the history of the system. Bohr points out that, for example, the length of the year in the Earth-Sun system is not determined by the masses of the Sun and the Earth but depends largely on the previous history all the way to the primordial phases of the formation of the solar system. If a large body would happen to traverse the solar system it would perturb the orbit and change the length of the year.

In the case of atoms, on the other hand, their properties remain unchanged even when subject to relatively large perturbations. In fact, if we wait a time long enough, the perturbed atom will return to its original state which depends *only* on the masses and charges of the elementary particles making up the atom itself.

The emission spectrum of atomic hydrogen of which fig. 4.5 is a very convincing example. In this spectrum we see many emission lines at very specific wavelengths measured with very high accuracy and precision. The wavelengths of the spectral lines are quite independent from the history, i.e. treatment, of the substance.

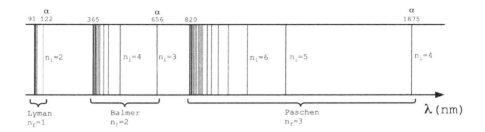

Figure 4.5 Lyman, Balmer and Paschen series in Hydrogen emission lines.

Another striking feature of the spectrum 4.5 consists in the evident regularities of the wavelengths at which the lines are located. It is quite natural to make the hypothesis that the electrons emit e.m. radiation as a result of a simple harmonic oscillation motion around some equilibrium position. Classical physics, however, cannot explain these spectra and new ideas are needed.

In order to account for the stability of the atom and the properties of the radiation emitted, Bohr made two simple assumptions: 1) Atomic systems exist in so-called *stationary states* where the charged particles obey the laws of classical mechanics but still possess mechanical stability, i.e. do not radiate e.m. radiation when in such a stationary state even if apparently subject to acceleration and therefore in violation of classical electrodynamics; 2) the emission of e.m. radiation is takes place if and only if the atom make a transition between two stationary states. Notice that a stationary states is defined also by the state where the energy of the system E is constant. Therefore, the process of emission of e.m. radiation between two stationary states characterized by energies E' and E'' is:

$$\Delta E = h\nu = E' - E'' \tag{4.2}$$

where h is Planck's constant and E' and E'' are, respectively, the initial and final stationary states. Obviously, in order to have emission, we must have $E' > E''$.

Bohr's postulates give immediately an explanation of Rydberg-Ritz combination principle through equation 4.2. Bohr was able to explain eq. 4.1 if we assume that the angular momentum L of an electron in his orbit around the proton is quantized:

$$L = m_e v r_n = n\hbar \tag{4.3}$$

where m_e is the electron mass, v its orbital velocity on the n-th circular orbit, with n an integer and $\hbar = \frac{h}{2\pi}$ is the reduced Planck constant. If we use classical mechanics, together with Bohr's postulates, we can not only derive the Rydberg formula 4.1, but also the value of the Rydberg constant R. In addition, a few more properties of the hydrogen atom like, for example the energy levels, the ionization energy and the size of electron orbits can be derived.

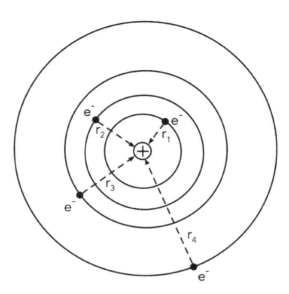

Figure 4.6 Bohr's model of the atom.

The size r_n of Bohr's orbits can be obtained by equating the centripetal force acting upon an electron with the electrostatic attraction:

$$\frac{m_e v_n^2}{r_n} = \frac{1}{4\pi\varepsilon_0} \frac{e^2}{r_n^2} \tag{4.4}$$

where $\varepsilon_0 = 8.854 \cdot 10^{-12}$ F/m is the vacuum permittivity and e the value of the elementary charge. Using eq. 4.3 we obtain the radius and the velocity of the electron in the nth orbit:

$$v_n = \frac{1}{4\pi\varepsilon_o} \frac{e^2}{n\hbar}$$
$$r_n = 4\pi\varepsilon_0 \frac{n^2\hbar^2}{m_e e^2} \tag{4.5}$$

Notice that in eq. 4.5 both the electron's velocity and the radius of the orbits depends only on the integer n, being all other quantities fundamental constants. We see that the size of the orbit is quadratic in n while the speed decrease linearly with n. If we insert $n = 1$ in the second equation in 4.5 we obtain the smallest orbit in the hydrogen atom. This radius is called *Bohr's radius* and is indicated as a_0:

$$a_0 = 4\pi\varepsilon_0 \frac{\hbar^2}{m_e e^2} \tag{4.6}$$

Bohr's radius is $a_0 = 5.29 \cdot 10^{-11}$ meters and is comparable with the measured size of the hydrogen atom. From eq. 4.5 we see that the speed of the electron in the first Bohr's orbit is $v_1 \sim 2 \times 10^6$ m/sec much smaller than the speed of light thus justifying the use of nonrelativistic mechanics.

If we keep applying classical mechanics to the orbiting electron, we can express its total energy as the sum of kinetic and potential energy:

$$E_n = T_n + V_n = \frac{1}{2}m_e v_n^2 - \frac{1}{4\pi\varepsilon_0}\frac{e^2}{r_n} \tag{4.7}$$

Inserting for v_n and r_n the values in eq. 4.5 we find:

$$E_n = \frac{1}{32\pi^2\varepsilon_0^2}\frac{m_e e^4}{\hbar^2}\frac{1}{n^2} - \frac{1}{16\pi^2\varepsilon_0^2}\frac{m_e e^4}{\hbar^2}\frac{1}{n^2} = -\frac{1}{32\pi^2\varepsilon_0^2}\frac{m_e e^4}{\hbar^2}\frac{1}{n^2} \tag{4.8}$$

and we see that the energy depends only on the index n, as in the case of the velocity and radius of the orbits. The energy of the first inner orbit of the hydrogen atom is:

$$E_1 = -\frac{1}{32\pi^2\varepsilon_0^2}\frac{m_e e^4}{\hbar^2} = -13.6 \text{ eV}. \tag{4.9}$$

Notice that the index "1" in the energy is referring to $n = 1$. Using eq. 4.9 we can rewrite the energy as:

$$E_n = \frac{E_1}{n^2} \tag{4.10}$$

which shows that there are an infinite number of energy states. More important, eq. 4.10 shows that the orbiting electron can only have e discrete set of energies, i.e. energy is *quantized*. There is only one energy state with the lowest energy possible, E_1, corresponding to $n = 1$, which is called *ground state*.

In complete analogy with a gravitationally bound body orbiting another body discussed in chapter 1, the negative energy of the electron means that the electron is bound to orbit the proton, with the circular orbit being the lowest energy.

The states with higher values of n are called *excited states*. We see that for $n \to \infty$ the energy 4.10 tends to zero, which corresponds to the electron not being bound to the proton anymore. As a consequence, we can interpret the energy E_1 as the minimum energy needed to extract an electron from the hydrogen atom, being E_1 the lowest energy level. This energy is called *ionization energy*. As a consequence, the bounded electron inside a hydrogen atom has its energy comprised in the interval $-13.6 < E < 0$ eV.

We can build the so called *energy spectrum* of the Hydrogen atom by plotting a series of horizontal lines on a vertical scale representing the set of stationary states (see fig. 4.7). The bottom line at energy $E_1 = -13.6$ eV is the ground state. The first state at $n = 2$ has an energy of $E_2 = -3.4$ eV and so on. Although discrete, there are an infinite number of negative energy states tending to zero energy, when the electron is not bound to the atom anymore. The predicted ionization energy of -13.6 eV is in good agreement with experimental data.

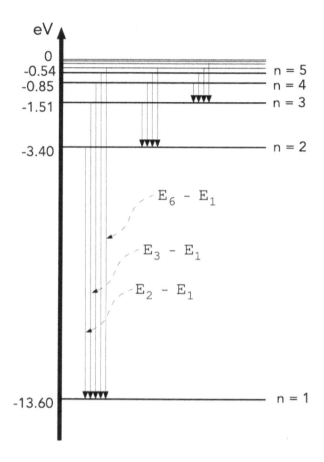

Figure 4.7 Hydrogen atom energy levels.

According to Bohr, the mechanism of emission of e.m. wave is represented by the downward vertical lines connecting states at different n. Photons are emitted only when an electron makes a transition from one stationary state to another stationary state. The energy of the emitted photon is then given by:

$$E_{nm} = h\nu = E_n - E_m = \frac{E_1}{n^2} - \frac{E_1}{m^2} \tag{4.11}$$

emission is possible only under the condition that $n > m$.

We now show the power of Bohr's model by deriving the Rydberg formula 4.1. If we express the frequency in terms of wavelength in eq. 4.11 we have:

$$\frac{1}{\lambda} = \frac{E_1}{hc}\left(\frac{1}{n^2} - \frac{1}{m^2}\right) \tag{4.12}$$

By comparing eq. 4.1 with eq. 4.12 we see that Bohr's theory is capable of expressing the Rydberg empirical constant R_H in terms of fundamental constants. In fact, it turns out that the Bohr's value for Rydberg constant:

$$\frac{E_1}{hc} = \frac{1}{32\pi^2\varepsilon_0^2}\frac{m_e e^4}{\hbar^2}\frac{1}{hc} = 1.097 \cdot 10^{-2}\text{nm}^{-1} \tag{4.13}$$

in excellent agreement with the experimental values.

Absorption is also easily explained by reversing all the arrows in fig. 4.7. In this case, when the n-th energy level is lower than the m-th energy level, then an electron jumps from the n-th to the m-th orbit by absorbing a photon with energy exactly equal to $E = E_m - E_n$. Absorption spectra then looks line black lines in a continuous spectrum.

4.1   DE BROGLIE

We have seen that Bohr's model, based on a few assumptions, is capable of explaining some peculiar features of the hydrogen atom. The two main successes are: explanation of Rydberg formula and expression of Rydberg empirical constant in terms of fundamental constants and the calculation of the size of the atom in agreement with measurements. The stability of the atom was explained by simply stating that electrons exist in stationary orbits and, for unknown reasons, do not emit e.m. waves even if subject to a constant acceleration. Let's remind us that, according to classical electromagnetism, accelerated charges radiate. Therefore, an electron orbiting around a proton should emit e.m. waves, lose energy and spiral rapidly toward the proton making matter highly unstable. According to classical physics, matter is inherently unstable.

In an attempt to understand the dynamics of the orbiting electrons, Louis de Broglie[2] made an assumption in his PhD thesis [11] in 1924: electrons, and elementary particles more in general, are characterized by a combination of matter and wave properties. To each elementary particle, there is a wave associated which is needed to determine its dynamical properties. This concept is referred to as *wave-particle duality*.

Of such a duality we had a glimpse when we discussed Einstein's photon hypothesis. In fact, we have seen that the photon has particle properties when measuring the Compton scattering while having also wave properties as demonstrated by light interference and diffraction. De Broglie, intuition consisted in extending this duality of light to matter particles.

De Broglie started with Einstein's special relativity mass-energy relation:

$$E = m_0 c^2 \tag{4.14}$$

---

[2]Louis Victor Pierre Raymond, 7th Duc de Broglie (1892-1987) was a French physicist who made important contributions to the development of quantum theory as described in the text. He was awarded the Nobel prize for physics in 1929.

where m_0 is the rest mass of a particle. De Broglie now used Planck's relationship:

$$E = h\nu \tag{4.15}$$

and equating eq. 4.14 with eq. 4.15 we have:

$$h\nu = m_0 c^2 \tag{4.16}$$

or in terms of wavelengths:

$$\frac{h}{\lambda} = m_0 c \tag{4.17}$$

Eq. 4.17 seems to hint that there is a relationship between the momentum of a particle (the right-hand side of 4.17 is the product of the mass with a velocity) with some wave-like property of wavelength λ, associated to the particle itself. De Broglie called these waves *Pilot Waves*. This wavelength is not to be confused with the wavelength of e.m. radiation. De Broglie, therefore, made the bold assumption that, to each massive particle, there is an associated wave of wavelength λ given by:

$$\lambda = \frac{h}{p} \tag{4.18}$$

where p is the momentum of a particle of mass m.

If we now allow this wave to interfere with itself we discover something interesting: the stable orbits of the hydrogen atoms, successfully calculated by Bohr, are those orbits for which the pilot waves are stationary, i.e. are standing waves. In fact, from eq. 4.4 we have:

$$e^2 = 4\pi\varepsilon_0 m_e v^2 r_n \tag{4.19}$$

and if we plug the value for e^2 from eq. 4.19 into r_n in eq. 4.5 we have:

$$2\pi r_n = \frac{h}{mv} \cdot n = n\lambda \tag{4.20}$$

which tells us that Bohr's quantized orbits are those orbits for which de Broglie pilot waves are stationary and a multiple of the integer n. We can visualize de Broglie ideas by wrapping a stationary 2D wave into itself on a series of concentric circles as shown in fig. 4.8.

De Broglie hydrogen atom, as depicted in fig. 4.8 resembles a musical instrument where the lowest energy orbit, corresponding to $n = 1$, is the fundamental tone and the highest energy orbits, corresponding to higher values of n, are the overtones. The condition in eq. 4.20 corresponds to having one full wavelength of the pilot wave when $n = 1$ with two nodes of the stationary wave. The higher n orbits are characterized by stationary waves with $2n$ nodes. De Broglie hydrogen atom model is mathematically equivalent to Bohr's model but it adds the idea, as pointed out by Gamow [17], that the electron orbiting the proton is characterized by an associated wave, of wavelength $\lambda = \frac{h}{p}$, capable of interfering with itself. The fact that this associated wave is stationary is supposed to be related to the fact that the electron does not spiral down into the proton emitting e.m. waves and thus making matter stable.

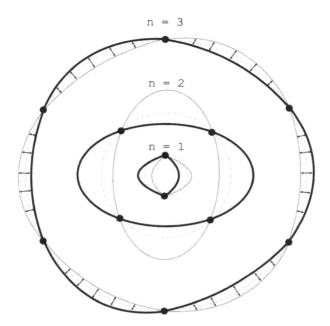

Figure 4.8 De Broglie hydrogen atom model. The first three circular orbits are shown together with the first three pilot wave standing waves. The black dots are the nodes of the standing waves.

4.2 DAVISSON AND GERMER EXPERIMENT

The remarkable fact that de Broglie hypothesis of pilot waves, when applied to the hydrogen atom, gives excellent agreement with Bohr's theory seems to give foundation that the matter-wave duality is a characteristic of elementary particles like the electrons. Shortly after de Broglie suggestion an experimental verification of de Broglie ideas was successfully attempted.

In 1927 two American physicists, Davisson[3] and Germer[4], devised the experiment sketched in fig. 4.9 [10]. It was known that diffraction is generated when X-rays impinge on a crystal. In particular, the angle of maximum reflection is given by the Bragg's law:

$$n\lambda = 2d \sin \theta \qquad (4.21)$$

where d is the spacing of the crystalline planes and θ is the angle between the incident rays and the crystalline plane responsible for the diffraction. In complete

---

[3]Clinton Joseph Davisson (1881-1958) was an American physicist best known for his experiment on the diffraction of electrons described in the text. He won the Nobel prize for physics in 1937.

[4]Lester Halbert Germer (1896-1971) was an American physicist who, with Clinton Davisson, proved that electrons diffract according to de Broglie hypothesis.

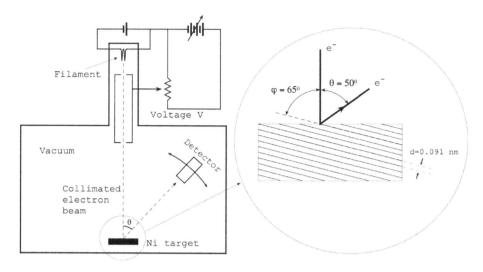

Figure 4.9 Davisson and Germer experiment. A collimated beam of electrons is focussed on a Nickel target. Diffracted electrons are recorded at various angles θ.

analogy with Bragg's diffraction of X-rays, if we accelerate electrons to the proper kinetic energy, with a corresponding momentum, then we can match the X-rays wavelength and expect electrons diffracting with the same law given by 4.21. The only difference will be that, instead of e.m. radiation of wavelength λ, we now have electrons with associated a pilot wave with the same wavelength λ.

Davisson and Germer needed to choose a crystal whose spacing was such that, under a reasonable acceleration voltage of the electrons, the de Broglie wavelength would be of the order of the lattice spacing.

It turned out that Nickel was a good candidate because it was already studied with conventional X-ray diffraction and so its lattice geometry was well understood. It turned out that the Nickel crystal had a spacing between atom planes of 0.091 nm.

Davisson and Gemer prepared their Nickel sample in such a way that the perpendicularly incident beam of electrons interacted with the lattice as shown in fig. 4.9. When they run their experiment, it was found that the electron detector measured a clear peak in the reflected electrons at an angle of 50° when they accelerated the electrons to a potential difference of $V = 54$ Volts, as shown in fig. 4.10. If we now calculate, as Davisson and Germer did, the de Broglie wavelength associated with the electrons we find that:

$$\lambda_{dB} = \frac{h}{p_e} \simeq \frac{h}{\sqrt{2m_e T}} \qquad (4.22)$$

where λ_{dB} is the de Broglie wavelength and p_e is the momentum of the electrons found using the nonrelativistic definition of kinetic energy $T = \frac{1}{2}mv^2$, from which

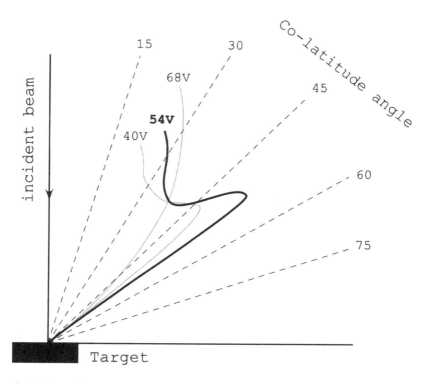

Figure 4.10 Intensity of scattered electrons vs the incident angle of the collimated electron beam to the Ni target. A clear peak is shown when the electrons are accelerated at a voltage $V = 54$ V.

$mv = \sqrt{2mT}$. Eq. 4.22 is valid for nonrelativistic electrons, i.e. for electron speeds much less than the speed of light[5], $v \ll c$.

Electrons, when subject to a potential difference of V_0 volts, acquire a kinetic energy equal to V_0 eV or, $V_0 \times 1.602 \cdot 10^{-19}$ J. Using for the electron mass the value $m_e = 9.31 \cdot 10^{-31}$ kg, eq. 4.22 becomes:

$$\lambda_{dB} \simeq \frac{h}{\sqrt{2m_e T}} = 0.166 \text{ nm} \qquad (4.23)$$

Now, let's verify that using Bragg's equation 4.21, we obtain the same wavelength for the e.m. waves. An inspection of fig. 4.9 shows that the angle of incidence to be used is $\varphi = 65°$ because the plane of the atoms responsible for the diffraction is inclined with respect to the face of the crystal normal to the incident electrons beam.

---

[5]In fact, the kinetic energy of the electron of 54 eV is negligible with respect to the rest mass of the electron of 0.51 MeV.

Using an angle of 65° in eq. 4.21, we have:

$$\lambda_{em} \simeq 2 \times 0.091 \times \sin 65° = 0.166 \text{ nm} \tag{4.24}$$

in excellent agreement with eq. 4.23. Notice that we used $n = 1$ in eq. 4.21.

The agreement between the results in eqs. 4.23 and 4.24 indicates strongly that the de Broglie pilot waves hypothesis is capable of explaining a number of experiments where classical physics fails. It would be very difficult to explain such behavior if we consider electrons as small classical particles: what is interfering?

4.3 THE RELEVANCE OF PLANCK'S CONSTANT

We have seen that, in order to explain quite a number of experimental results, Planck had to introduce the concept of quantization of energy, according to eq. 3.31. The fact that Planck's constant h is small, being of the order of 10^{-34} J·s, gives us an explanation of why certain physical phenomena seem to be relegated to systems that are very small compared to the size of human beings. Notice also that Planck's constant has the physical dimension of mechanical action, i.e. energy × time or momentum × length.

In order to set the scale of where (and when) quantum phenomena become important, we need to discuss Planck's constant in more detail. The first consideration to be made consists in noticing that every time a new fundamental constant, i.e. a constant that is not expressed in term of other constants but needs to be measured, opens a new field of physics. Newton's gravitational constant G, for example, opened the field of gravity – first with Newton and then with Einstein's general relativity theory. The relatively simple fact that the speed of light is a constant, and that it does not depend on the state of motion of the observer has generated Einstein's special relativity theory. Planck's constant, as we are discussing in this book, is intimately linked with quantum mechanics. The units of h are J·s which, in classical physics, are the unit of *action*.

We have already encountered the concept of mechanical action when we discussed the Lagrangian formulation of classical mechanics. We have stated, as a principle, that the equation of motion of a mechanical system can be derived through the principle of stationary action. This elegant and powerful formulation of classical mechanics tells us that the concept of action plays a fundamental role. The concept of action, through Planck's constant, comes back also when a new kind of theory is needed to explain phenomena that cannot be explained with classical physics. Now there is a big and important new concept: Planck's constant tells us that the action is quantized and that there is a **minimum** amount of action in any mechanical system. Nature is engineered in such a way that, no matter the details of any mechanical system, the amount of action associated with it cannot be less than h.

According to Maupertuis' principle, already briefly discussed in Chapter 1, the action is defined as:

$$A = \int_{q_1}^{q_2} p\,dq \tag{4.25}$$

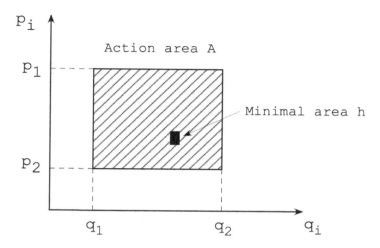

Figure 4.11   Planck's constant as minimal area in the phase space.

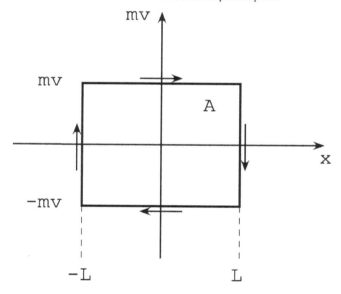

Figure 4.12   Phase space diagram of a tennis ball of mass m bouncing back and forth between the baselines with momentum mv. The length of the tennis court is $2L$.

This means that the numerical value of the Planck's constant represents a small area in the phase space p, q as shown in fig. 4.11. The large rectangle represents a system bound between the two coordinates q_1, q_2 and the two momenta p_1, p_2. The action A is then the area $A = (q_2 - q_1) \cdot (p_2 - p_1)$. Quantum mechanics is important when $A \sim h$. When we say that h is small, we mean that is small with respect to our everyday experience. For example, the action associated with the motion of a tennis ball during a tennis match can be estimated by looking at the phase diagram of fig. 4.12.

For a tennis court of length $2L \simeq 24$ m, with a ball of mass $m \simeq 58$ g traveling at a speed of $v = 55$ m/sec, the action is the area $A = 4mvL \simeq 158.5$ J $\cdot$ s. This value is 36 orders of magnitude larger than Planck's constant which means that we can safely neglect quantum effects when playing tennis.

Let us now consider the hydrogen atom and verify that the action associated with the electron bound to the proton has an action comparable with the Planck's constant. Eq. 4.25, in the case of a circular orbit as is the case of the first orbit when $n = 1$ in eq. 4.8, is:

$$A = \oint pdq = 2\pi r_1 p_1 \tag{4.26}$$

where the integral is calculated over the first full orbit and r_1 and p_1 are, respectively, the radius and the momentum of the electron in the first orbit. Eq. 4.5 gives us the numerical values for r_1 and p_1 that can be inserted in eq. 4.26 to give $A \simeq 10^{-34}$ J·s which is of the same order of magnitude of the Planck's constant. Therefore, when dealing with elementary particles, like the electrons, bound into systems whose dimensions are of the order of atoms or molecules, we conclude that we must use quantum mechanics. Notice that the orbital speed of the electron, when calculated using eq. 4.5, results to be of the order of 10^6 m/s much less than the speed of light thus allowing us to use the nonrelativistic expression of momentum $p = mv$.

5 Schrödinger Equation

In this chapter, we introduce the Hamilton-Jacobi equation showing that classical mechanics can be described with a wave equation for the classical action. We then discuss the double-slit experiment, first with e.m. waves and then with massive particles like, for example, electrons. This will allow us to introduce the concept of probability amplitudes and their role in interference phenomena. We will then show how particles can be subject to interference and diffraction phenomena. Uncertainty, principle will be then deduced using a mixture of classical and quantum ideas. Finally, the Schrödinger equation is introduced using Feynman path integral formulation of quantum mechanics.

5.1 HAMILTON-JACOBI

We have seen in the previous chapters that classical mechanics was not able to explain a number of experimental results obtained in many different circumstances. From these experiments, it was evident that, when we probe the world of small particles, i.e. when the action is of the order of Planck constant, a new physics is needed. This new physics should be able to incorporate a few basic and important facts. First, certain physical quantities come in quanta like, for example, the energy exchange between matter and e.m. radiation in a blackbody cavity, or the energy of photons. Second, the happening of some physical phenomena like, for example, the timing of the radioactive decay of an unstable nucleus, cannot be predicted even in principle. This implies that some level of probabilistic interpretation needs to be incorporated into the description of physical phenomena. Third, e.m. radiation as well as matter particles seem to have, at the same time, wave-like and particle-like characteristics. There are experimental conditions where particles, like electrons, behave as waves, and there are other experimental conditions where e.m. waves behave as particles.

The distinction between the wave-like and particle-like description of certain physical phenomena relies almost entirely on our idealization of the physical world. The fact that we *see* material objects precisely localized in space, leads us to imagine that such precise localization is still valid if the particles become smaller and smaller. The advent of more and more powerful microscopes, for example, has shown that tiny material particles, down to almost atomic scale, can still be observed localized in space. It is quite natural, therefore, to imagine that, no matter how small the particle is, it can still be observed spatially localized with as high accuracy as we like. Therefore it does not surprise that Newtonian abstraction of a particle consists in having a finite mass characterized by zero spatial extension and perfectly localized in a point in space[1].

---

[1] Of course, this is a highly ideal situation. Einstein's general relativity tells us that such a particle would collapse into a black hole.

DOI: 10.1201/9781003145561-5

On the other hand, we have seen that, in order to explain diffraction and interference of small particles, like electrons, we need to associate a wave whose origin and characteristics are still to be precisely determined. The stability of the atom itself, as we have seen, seems to require that a wave capable of interfering with itself must be associated with the motion of electrons around the nucleus. Both Lagrangian and Hamiltonian formalism of classical mechanics, reviewed in a previous chapter, *apparently* seem not to allow having wave-like and particle-like attributes to a point mass. However, there is a third formalism, based on Hamilton-Jacobi equation, that establishes a direct link between matter and waves. We will now show that, contrary to common sense, classical mechanics already contains, to some extent, the duality wave-particle.

5.1.1 CANONICAL TRANSFORMATIONS

It is interesting, and probably surprising to our classical minds, that classical mechanics can be described by a wave equation. In order to find the equation showing that classical mechanics already contains a wave behavior in matter particles, we need to go back to the Lagrange/Hamilton formalism. In particular we will now introduce a special class of generalized coordinate transformations called *canonical transformations*.

We have seen already coordinate transformations from a set of generalized coordinate q to a new set of generalized coordinates q'. We have operated such transformations either to simplify the physical problem that we are trying to describe or, at least, to try to make the problems more intuitive. An example is the change from Cartesian (x, y) to polar (r, θ) when dealing with central forces or rotating frames of reference.

We now want to be more general and consider transformations that involve coordinates and momenta at the same time:

$$q' = q'(q, p, t)$$
$$p' = p'(q, p, t) \tag{5.1}$$

Let's assume that the unprimed coordinates satisfy Hamilton's equations 1.80, that we repeat here for convenience:

$$\dot{q} = \frac{\partial H}{\partial p}$$

$$\dot{p} = -\frac{\partial H}{\partial q} \tag{5.2}$$

$$-\frac{\partial L}{\partial t} = \frac{\partial H}{\partial t}$$

where an index i running from 1 to N has been omitted for clarity. Those transformations of the form 5.1 for which the new primed coordinates also satisfy Hamilton's equations 5.2 are called *canonical transformations*.

Let's now recall from chapter 1 the Legendre transform that we used to define the Hamiltonian:

$$H(q,p) = \sum \dot{q}p - \mathscr{L} \tag{5.3}$$

from which we obtain the Lagrangian $\mathscr{L} = \sum \dot{q}p - H(q,p)$. We can now write a *minimum action* condition, in analogy to eq. 1.14:

$$\delta \int_{t_1}^{t_2} \mathscr{L}\,dt = \delta \int_{t_1}^{t_2} (\sum \dot{q}p - H(q,p))dt = 0 \tag{5.4}$$

If the transformation 5.1 is a canonical transformation, then we can write eq. 5.4 also for the transformed coordinates q', p':

$$\delta \int_{t_1}^{t_2} (\sum \dot{q}'p' - H'(q',p'))dt = 0 \tag{5.5}$$

In order to satisfy simultaneously eqs. 5.4 and 5.5, the two integrands should be equated and they can differ at most by the total derivative of an arbitrary function F such that:

$$\int_{t_1}^{t_2} \frac{dF}{dt}\,dt = F(t_2) - F(t_1). \tag{5.6}$$

The condition 5.6 automatically insures that the variations 5.4 and 5.5 are zero because the value of the integral 5.6 does not depend on the path chosen depending only on the fixed extremes at t_1 and t_2. The function F is called generating function of the canonical transformation. In fact, knowing F determines the equations of transformations 5.1.

The generating function F is in general a function of the unprimed and primed coordinate plus the time (q, p, q', p', t) for a total of $4N$ variables. However, the transformations 5.1 reduce the total number of independent variables to $2N + 1$. Therefore F can be one of the following: $F_1(q,q',t)$, $F_2(q,p',t)$, $F_3(p,q',t)$ and $F_4(p,p',t)$. The choice of the generating function is arbitrary and usually dictated by convenience when studying a specific problem.

If we choose the form $F_1 = F_1(q,q',t)$ the condition for canonical transformation is:

$$\sum \dot{q}p - H(q,p) = \sum \dot{q}'p' - H'(q',p') + \frac{dF_1(q,q',t)}{dt} \tag{5.7}$$

We can write the total derivative of F_1 as:

$$\frac{dF_1}{dt} = \sum \frac{\partial F_1}{\partial q}\dot{q} + \sum \frac{\partial F_1}{\partial q'}\dot{q}' + \frac{\partial F_1}{\partial t} \tag{5.8}$$

which inserted in eq. 5.7 gives:

$$\sum \dot{q}p - H(q,p) = \sum \dot{q}'p' - H'(q',p') + \sum \frac{\partial F_1}{\partial q}\dot{q} + \sum \frac{\partial F_1}{\partial q'}\dot{q}' + \frac{\partial F_1}{\partial t} \tag{5.9}$$

Eq. 5.9 is satisfied if and only if the coefficients of the independent variables $\dot{q}$ and $\dot{q}'$ are separately equal to zero. This is true if:

$$p = \frac{\partial F_1}{\partial q}$$

$$p' = -\frac{\partial F_1}{\partial q'} \qquad (5.10)$$

$$H' = H + \frac{\partial F_1}{\partial t}$$

If instead of using F_1 we were to choose $F_2 = F_2(q, p', t)$ we would need to go from the variables (q, q') to the variables (q, p'). This is achieved by imposing the following Legendre transform:

$$F_2(q, p', t) = F_1(q, q', t) + \sum p'q' \qquad (5.11)$$

For simplicity we call $S = F_2$ so that eq. 5.11 can be written as:

$$\sum \dot{q}p - H(q, p) = \sum \dot{q}'p' - H'(q', p') + \frac{d}{dt}[S(q, p', t) - \sum q'p']$$

$$= \sum \dot{q}'p' - H'(q', p') + \frac{dS(q, p', t)}{dt} - \sum \dot{q}'p' - \sum q'\dot{p}' \qquad (5.12)$$

$$= -\sum q'\dot{p}' - H'(q', p') + \frac{dS(q, p', t)}{dt}$$

Repeating the same substitution as in the previous case for F_1 we have:

$$p = \frac{\partial S}{\partial q}$$

$$q' = -\frac{\partial S}{\partial p'} \qquad (5.13)$$

$$H' = H + \frac{\partial S}{\partial t}$$

5.1.2 HAMILTON-JACOBI EQUATION

Let us now search for a very special canonical transformation from the original (q, p) coordinates to a new set of coordinates $q' = $ constant and $p' = $ constant. Let us also specify the value of these constants to be the $2N$ initial conditions (q_0, p_0). If we can find such a canonical transformation then the equations relating the old to the new variables is exactly the solution to the problem:

$$q = q(q_0, p_0, t)$$

$$p = p(q_0, p_0, t) \qquad (5.14)$$

It is easy to verify that if we impose that the new Hamiltonian H' is identically equal to zero, then the new variables (q', p') are constant. In fact, Hamilton's equations for the transformed coordinates are:

$$\dot{q}' = \frac{\partial H'}{\partial p'} = 0$$
$$\dot{p}' = -\frac{\partial H'}{\partial q'} = 0 \tag{5.15}$$

from which it follows 5.14. The third equation in 5.13, with the condition $H' = 0$ becomes:

$$H(q,p,t) + \frac{\partial S}{\partial t} = 0 \tag{5.16}$$

and using the first equation in 5.13, we can formally rewrite eq. 5.16 as:

$$H(q, \frac{\partial S}{\partial q}, t) + \frac{\partial S}{\partial t} = 0 \tag{5.17}$$

which is the *Hamilton-Jacobi* (HJ) equation. The HJ equation is a first-order partial differential equation for the $N + 1$ variables (q,t). Its solution S is called *Hamilton's principal function*.

Knowing the function S, solution of the HJ equation, is equivalent to solve the mechanical problem as we will see later with a simple example. We have seen that imposing $H' = 0$ is equivalent to transform to a new coordinate system where the generalized coordinates q' and generalized momenta p' are constant. This means that the function S depends only on q and t, i.e. $S = S(q,t)$. Using eq. 5.8, the total derivative of the function $S = S(q,t)$ becomes:

$$\frac{dS}{dt} = \sum \frac{\partial S}{\partial q}\dot{q} + \frac{\partial S}{\partial t} \tag{5.18}$$

using eq. 5.10 we have:

$$\frac{dS}{dt} = \sum p\dot{q} - H = \mathscr{L} \tag{5.19}$$

that integrated gives:

$$S = \int \mathscr{L}dt + \text{const.} \tag{5.20}$$

which is the *action* as defined in eq. 1.14 plus an arbitrary constant.

There is a more intuitive derivation [41] of the HJ equation that shows directly the connection with the action S. Let us write down again action 5.20 as an integral of the Lagrangian in the interval (t_0, t):

$$S = \int_{t_0}^{t} \mathscr{L}(q, \dot{q}, t)dt \tag{5.21}$$

Let us use the expression $\mathscr{L} = p\dot{q} - H$, where $p = \frac{\partial \mathscr{L}}{\partial \dot{q}}$. For a given path $q = q(t)$, we can write the action as:

$$
\begin{aligned}
S(q,t) &= \int_{t_0}^{t} p\dot{q}\,dt - \int_{t_0}^{t} H(p,q,t)\,dt \\
&= \int_{q_0}^{q} p\,dq - \int_{t_0}^{t} H(p,q,t)\,dt
\end{aligned}
\tag{5.22}
$$

Notice that eq. 5.22 has transformed a line integral over time, from t_0 to t, to a line integral over the plane q,t from the point (q_0,t_0) to the point (q,t). In general, the path integral of any function $S = S(q,t)$ can be written as:

$$
S(q,t) = \int_{q_0}^{q} \frac{\partial S}{\partial q}\,dq + \int_{t_0}^{t} \frac{\partial S}{\partial t}\,dt
\tag{5.23}
$$

From comparing eq. 5.22 and eq. 5.23 it follows that:

$$
\begin{aligned}
p &= \frac{\partial S}{\partial q} \\
\frac{\partial S}{\partial t} &= -H\left(\frac{\partial S}{\partial q},q,t\right)
\end{aligned}
\tag{5.24}
$$

which is exactly the HJ equation 5.17.

If the Hamiltonian H does not explicitly depend on the time t, energy is conserved and the second equation in eq. 5.24 becomes:

$$
H\left(\frac{\partial S}{\partial q},q\right) + \frac{\partial S}{\partial t} = 0
\tag{5.25}
$$

where we see that only the last term depends explicitly on time. We can therefore write a solution to eq. 5.25 in the form:

$$
S(q,p',t) = W(q,p') - Et
\tag{5.26}
$$

where E is a constant that we identify as the conserved energy. Due to the way in which we built the HJ equation, i.e. by requiring that q' and p' are constant, the action S has the form:

$$
S = S(q,\eta,t)
\tag{5.27}
$$

where $\eta = (\eta_1,\eta_2,...,\eta_N)$ is a vector of N independent integration constants. A good choice for these constants would be new momenta $p' = \eta$.

The function $W(q,p')$ is called *Hamilton's characteristic function*, and it does not depend on time. Eq. 5.26 tells us that the dynamics are represented by a surface moving with constant speed E in the space (q,p'). Pushing further the analogy, the function S represents the motion of the points of constant phase of a wave propagating with speed E. In fig. 5.1 we see the motion of the equi-phase surface $W(q_1)$ at time t_1 moving to $W(q_2)$ at time t_2 and then $W(q_3)$ at time t_3. Looking at the first

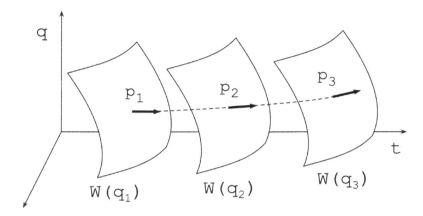

Figure 5.1 Motion of the S surfaces in the q space.

equation in eq. 5.13, it is easy to verify that the momentum p is always perpendicular to the surface W. We have therefore reached the conclusion that the dynamics of a classical mechanical system can be described by a *nonlinear* partial differential equation – the HJ equation – for the function S. Once we find the function S, the equation of motion are found by writing:

$$q'_n = \frac{\partial S(q_n, p'_n, t)}{\partial p'_n} \tag{5.28}$$

written explicitly for the $N+1$ components, where $n = 1, ..., N$ are the degrees of freedom plus the time t. The equation of motions are then obtained by solving eq. 5.28 for q_n:

$$q_n = q(q', p', t). \tag{5.29}$$

Similarly, we find the solution for p_n by writing:

$$p_n = \frac{\partial S(q_n, p'_n, t)}{\partial q_n}. \tag{5.30}$$

Let's go back to HJ equation and write down its form for a nonrelativistic particle ($v \ll c$) of mass m in a potential $V(r)$. The recipe is simple: write the Hamiltonian:

$$H = \frac{p^2}{2m} + V(r) \tag{5.31}$$

and substitute for p its expression $p = \frac{\partial S}{\partial q}$. In Cartesian coordinates, we have:

$$\frac{1}{2m}\left[\left(\frac{\partial S}{\partial x}\right)^2 + \left(\frac{\partial S}{\partial y}\right)^2 + \left(\frac{\partial S}{\partial z}\right)^2\right] + V(x, y, z) + \frac{\partial S}{\partial t} = 0 \tag{5.32}$$

which is a useful form of HJ equation involving the action S.

5.2 FEYNMAN PATH INTEGRALS

We have seen, especially in Chapter 3, that a significant number of experiments were giving results that could not be explained by classical physics. Probably the most important was the determination of the spectral character of the blackbody emission. In order to obtain a mathematical description of the blackbody spectrum Planck needed to make a basic assumption that energy is exchanged in quanta $E = h\nu$ between radiation and the oscillators in the walls of the blackbody. Einstein then extended the idea and proposed that light itself is quantized and not just in the energy exchange with oscillators, giving rise to the idea of photons as quanta of e.m. radiation. The photoelectric effect showed that photons behave as particles or waves depending on the specifics of the experimental setup thus hinting at the duality particle-waves. Photons have zero rest mass and De Broglie made the bold proposal that such duality is valid also for massive particles with mass $m \neq 0$. The discovery of radio activity added another piece of information to the birth of a new physics: there are physical phenomena that cannot be predicted with certainty but that seem to obey the law of probability. Radio activity introduces the concept of probability in our physical understanding of nature: it is not possible to foresee when a specific atom will emit a sub-atomic particle. Only observing a large number of identical systems we can calculate the probability of decay of a single atom.

Davisson-Germer experiment has shown without any doubt that electrons behave as massive particles whose dynamics is governed by some sort of wave. We therefore need a wave equation for a mathematical function that is able to explain all the known experimental facts and, if the theory is good, can predict new phenomena. We have seen that the action S with all the associated stationary principles seem to be playing a central role in the wave description of classical physics through its role in the HJ equation. The question is, can we modify the HJ equation to incorporate the new phenomena? The HJ equation, in fact, contains the action S and describes classical mechanics with a wave description where the action S plays a central role.

The route to a quantum mechanical analogue - the Schrödinger equation - of the HJ equation takes us first to analyze diffraction/interference experiments. These in turn will pave the road to the Feynman path integral formalism from which the Schrödinger equation follows naturally.

5.2.1 THE DOUBLE-SLIT EXPERIMENT FOR LIGHT

In a typical Young experiment, a light source of monochromatic waves of wavelength λ, with associated wave vector $k = \frac{2\pi}{\lambda}$, is illuminating a screen with two small slits as shown in fig. 5.2. We want to study the distribution of light intensity $I(\theta)$ on another screen - the observing screen - positioned behind the first screen at a distance $L = \overline{OO'}$.

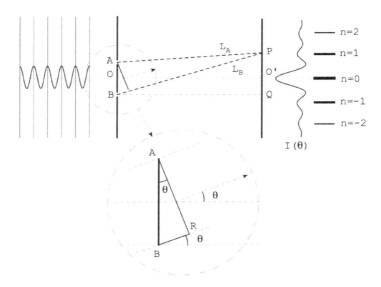

Figure 5.2 Interference of light. Interference by monochromatic light incident on a screen with two small holes at A and B. An interference fringe pattern $I(\theta)$ is observed. The location of the maxima are obtained when the waves along the paths L_A and L_B differ by an integer number n of wavelengths.

According to Huygens's principle[2] in order to have constructive interference at a point P on the screen, the two paths L_A and L_B must differ by an integer number of wavelengths, i.e. $L_B - L_A = n\lambda$. Simple geometry shows that, if θ is the angle $\angle QBP$ and if the second screen is positioned at a distance $\overline{OO'} \gg \overline{AB}$, then the path difference $\overline{BR} \approx \overline{AB}\sin\theta = d\sin\theta$, where $d = \overline{AB}$ is the distance between the two holes A, B in the screen. The condition of maximum interference is then:

$$d\sin\theta = n\lambda \tag{5.33}$$

For the first oder, $n = 1$, it is easy to verify that the distances of the maximas on the observing screen is:

$$\overline{O'P} = \overline{OO'}\tan\theta. \tag{5.34}$$

Similar formula can be obtained for higher orders $n > 1$.

We can describe a linearly polarized incident plane wave, propagating along the direction identified by the vector r, by writing the electric field E as:

$$E(r,t) = E_0 e^{i(\omega t - k \cdot r + \phi_0)} \tag{5.35}$$

---

[2]Huygens' principle states that every point on a wave front may be considered as a source of spherical waves.

We introduced the complex field $E(r,t)$ in eq. 5.35 as a practical way to calculate interference between two waves keeping in mind that only the real part has physical significance. In fact, when we compare with measurements, we have to consider only the real part of the field, i.e. $\text{Re}(E) = E_0 \cos(\omega t - k \cdot r + \phi_0)$. We can write explicitly the phase term of the incident e.m. wave by isolating the phase term $e^{i\phi_0}$ such that the e.m. field can be written as:

$$E(r,t) = e^{i\phi_0} E_0 e^{i(\omega t - k \cdot r)} = e^{i\phi_0} E_0(r,t) \tag{5.36}$$

The field at point P in fig. 5.2 is written as the sum of the two fields originated at the points A and B. We have:

$$E(P) = E(L_A, t) + E(L_B, t) \tag{5.37}$$

The intensity can be written as:

$$I(P) = |E(P)|^2 = |E(L_A, t) + E(L_B, t)|^2 \tag{5.38}$$

Eq. 5.38 can be written in terms of the angle θ as:

$$I(\theta) = |E_A + E_B|^2 = |E_A|^2 + |E_B|^2 + 2|E_A||E_B| \tag{5.39}$$

Eq. 5.39 tells us that the light intensity on the observing screen is the sum of the light intensity coming from the hole A plus the light intensity coming from the hole B plus the interference term $2|E_A||E_B|$.

In terms of the phase of the e.m. wave, the phase difference at the point P is:

$$\Delta\phi = 2\pi \frac{L_B - L_A}{\lambda} = 2\pi \frac{\Delta L}{\lambda}. \tag{5.40}$$

Fig. 5.3 shows a more geometrical – and perhaps intuitive – way to picture the interference between two e.m. waves.

It is important to stress that, in order to have interference, light needs to be effectively diffracted by the two holes in A and B. The condition that the pin holes A and B of fig. 5.3 must be smaller or of the order of the wavelength is needed to make sure that light is scattered at a large angle, almost spherically. In this way, light will illuminate the largest portion of the observing screen thus maximizing the portion of screen illuminated by both diffraction holes. If the hole is large compared to the wavelength, as in fig. 5.4 left panel, light will be less scattered and is concentrated mostly in the forward direction with less superposition of the two beams on the observing screen.

This can be intuitively understood by using Huygens' principle, according to which each point on a wave front can be thought as the source of a spherical wave. Left panel of fig. 5.4 shows that we can fit a relatively large number of sources of spherical waves. Since the separation between the sources is small compared to the wavelength, there will be constructive interference of the adjacent sources combining to produce an almost flat wave front. The larger the hole the flatter the emerging

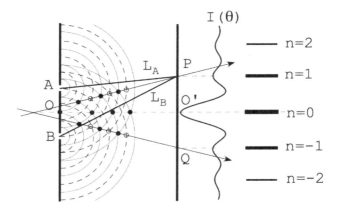

Figure 5.3 Geometric construction of the interference pattern. The filled circled points correspond to constructive interference while the open circled points correspond to destructive interference.

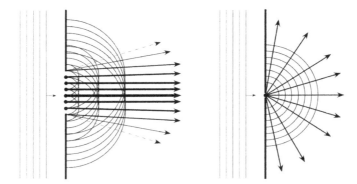

Figure 5.4 Diffraction of a plane wave incident on an aperture larger than the wavelength (left) compared to the diffraction on an aperture smaller or of the order of the wavelength. Small apertures diffract at large angle generating an almost spherical wave.

wave front. In the limit of an infinitely large hole, the plane wave propagates undisturbed. This is due to the fact that the phase difference between adjacent sources of spherical wave is very small and constructive interference is achieved in the forward direction. Right panel of fig. 5.4 instead shows that when less and less sources of spherical waves are present, light will spread more and more spherically. We can say qualitatively that the smaller we confine an incident plane wave through a hole, the larger will be the spread of light emerging from the hole.

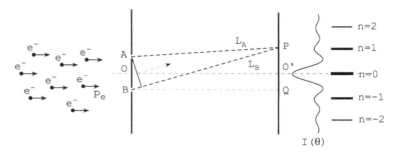

Figure 5.5 Interference by mono-energetic electrons incident on a screen with two small holes at A and B. An interference fringe pattern $I(\theta)$ is observed. The location of the maxima are obtained when the de Broglie waves along the paths L_A and L_B differ by an integer number n of wavelengths.

5.2.2 THE DOUBLE-SLIT EXPERIMENT FOR ELECTRONS

The logical step now is to make the hypothesis that, in complete analogy with the e.m. waves, a double slit experiment where particles like electrons are used, should produce interference effects of the associated de Broglie waves λ_{DB} with fringes generated by a path difference of the order of $\Delta\phi_{DB} = 2\pi \frac{\Delta L}{\lambda_{DB}}$.

In fig. 5.5, we have a beam of mono-energetic electrons, i.e. electrons all traveling with the same momentum p_e, directed on a screen with two holes with diameter of the order, or less, that the de Broglie wavelength $\lambda_{dB} = \frac{h}{p_e}$, where p_e is the momentum of the incident electrons. Electrons are detected at the observing screen, i.e. every time an electron arrives at the screen QP its position is detected and counted.

After enough electrons are counted as a function of the angle of arrival θ, or as a function of the position (x, y) on the observing screen, a typical interference plot $I(\theta)$ is observed (see fig. 5.5). To our classical mind, this result is somewhat surprising because if we think of electrons as just tiny massive balls, they can go through either slit A or slit B and interference cannot be observed. In fact, if we call $I(A)$ and $I(B)$, respectively, the distribution of electrons on the observing screen when alternatively the slit B or A is closed, we would expect $I(\theta) = I(A) + I(B)$ without any interference term. To have interference, we know that we need to interpret the intensity terms $I(\theta), I(A)$ and $I(B)$ as the square modulus of an amplitude. De Broglie waves seem to be the perfect wave to associate to the motion of the electrons.

We have seen already in chapter 4 that there was robust experimental evidence (Davisson-Germer) that electrons can be described as massive particles associated with a de Broglie wave responsible for producing interference. If we want to continue along the analogy with Young's experiment, we can try to use the de Broglie wave in analogy with the amplitude of the electric field so that we can use the mathematical machinery of e.m. interference already developed. Let's indicate the de Broglie wave[3] with the function $\psi(r,t)$ which is the analog of the electric field $E(r,t)$. If we now allow the functions ψ to interfere, we will have that the intensity recorded is:

$$I(\theta) = |\psi(A) + \psi(B)|^2 = |\psi(A)|^2 + |\psi(B)|^2 + 2|\psi(A)| \cdot |\psi(B)| \qquad (5.41)$$

where $\psi(A)$ and $\psi(B)$ are, respectively, the amplitudes of the functions associated with the path L_A and L_B. Interference is observed depending on the phases associated with the functions $\psi(A)$ and $\psi(B)$. It is clear that, as in the case of e.m. interference, the relative phase of the two functions is the important parameter to be studied if we want to describe the interference among electrons.

We are now in a position to associate a better physical meaning to the function ψ. If we let the electrons arrive one at a time well separated, it is impossible to predict with certainty where on the observing screen the electron will land. An interference figure will appear only after a substantial number of electrons are allowed to reach the observing screen and detected[4].

If we instead look at the path of a single electron at the time, we seem to be incapable of predicting uniquely its trajectory. The best we can do is to wait until the position of enough number of electrons is determined and observe their distribution on the observing screen. The crucial step consists in observing that the *distribution* of recorded electrons can be modeled by using the same maths/physics that we used for studying the interference of e.m. waves[5]. For now, it seems therefore quite natural to associate a *probabilistic* meaning to the wavefunctions. From now on we will refer to **wavefunctions** $\psi(r,t)$ the waves associated with the electrons that are responsible for the interference pattern $I(\theta)$ of fig. 5.5. In analogy with the interference of e.m. waves, we make the assumption that the electron distribution function observed at the point P, i.e. $I(P)$, is proportional to the square modulus of the wavefunction:

$$I(P) \propto |\psi(P)|^2. \qquad (5.42)$$

where $\psi(P)$ is the wavefunction calculated at the point P on the observing screen. If we normalize the wavefunction in such a way that its integral calculated over the entire observing screen is equal to one:

$$\int_S |\psi(x,y)|^2 dx dy = 1, \qquad (5.43)$$

where S is the area of the observed screen described by the two axis x and y, then $|\psi(x_0,y_0)|^2$ is exactly the probability of observing an electron at point P of coordinates (x_0,y_0). More in general, if we are not limited to a plane, the normalization

---

[3]This association is not completely correct because the quantum wave function is much more than just the de Broglie wave. We will discuss this point at length later in the book.

[4]Incidentally, this demonstrates also that electrons do not interfere with each other: interference is observed because the wave associated with each individual electron is split when impinges on the first screen so that it can interfere later on the observing screen.

[5]With some important differences that we will discuss later.

condition of the wavefunction is:

$$\int_V |\psi(x,y,z)|^2 dV = 1, \qquad (5.44)$$

where V is the volume in which the wavefunction is defined. This is the well-known Born[6] probabilistic interpretation of the wavefunction also known as **Born rule**. Notice that, since we are dealing with the square modulus, the most general assumption would be to have the wavefunction expressed as a complex function.

We are left with finding the equation for the wavefunction ψ. Before the correct equation was found, some attempts were made based on the analogy with the e.m. wave equation. Unfortunately (or fortunately) they all gave problems when confronted with experimental data. One way out is to go back to the basics and start with a new set of axioms. We discuss first the so called *path integral formulation of quantum mechanics* by Feynman. Once we obtain the correct equation[6] we will study different ways to "justify" it.

5.2.3 FROM PATH INTEGRALS TO SCHRÖDINGER EQUATION

We now introduce the Feynman's path integral formulation of quantum mechanics and show that the wave equation that we are after, i.e. the time-dependent Schrödinger equation, can be obtained from the HJ equation providing that we give also a set of axioms.

As with every physical theory, we need to establish and accept a set of basic postulates. There is no way to "prove" that postulates are right or wrong: scientists will accept the postulates as long as the theory built on them is explaining all known experimental facts. A good theory, like quantum theory, is also capable of predict new phenomena. Feynman's theory [15] is not an exception and is based on the following two postulates:

I If an ideal measurement is performed to determine whether a particle has a path lying in a region of spacetime, the probability that the result will be affirmative is the absolute square of a sum of complex contributions, one from each path in the region;

II The paths contribute equally in magnitude, but the phase of their contribution is the classical action (in units of $\hbar$), i.e. the time integral of the Lagrangian taken along the path.

Postulate I is probably inspired by the double slit experiment for electrons and is a statement about the *superposition principle* plus Born's probabilistic interpretation of the wavefunctions. Postulate II gives the recipe for calculating the amplitude of each possible paths.

It is critical to stress, at this point, that all possible paths are calculated between an initial point $P_i(x_i, t_i)$ and a final point $P_f(x_f, t_f)$. This means that all the paths are traveled during the same time interval $\Delta t = t_f - t_i$.

---

[6]Schrödinger equation is correct when relativistic effects are negligible.

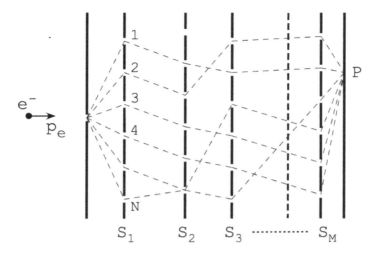

Figure 5.6   Interference from a set of M screens containing each N slits.

Let's turn our attention to postulate I and let's make a more complex version of the double slit experiment for mono-energetic electrons. For simplicity we consider the diffraction from a single slit and we imagine that we position between the source of electrons and the observing screen a series of M screens $S_1, S_2, ..., S_M$ each containing N slits as depicted in fig. 5.6. If we impose the superposition principle, then the amplitude at point P will be the sum of all possible amplitudes associated with all possible paths, according to:

$$\phi(P) = \sum_{i=1}^{N} \sum_{k=1}^{M} \psi(x_k^i). \tag{5.45}$$

In the limit for $M, N \to \infty$, the screens and the slits effectively disappear and we have now propagation in empty space, as shown in fig. 5.7. We are now in the position to express the probability amplitude for a particle moving from an initial position (x_i, t_i), i.e. the particle located at x_i at the time t_i, to a final position (x_f, t_f) where the particle is located at x_f at time t_f.

Postulate II tells us that each path contributes an amplitude of the form:

$$\phi^j = F_{fi}^j e^{\frac{i}{\hbar} S_{fi}(x^j)} \tag{5.46}$$

where F_{fi}^j are real functions, x^j is the jth path and:

$$S_{fi} = \int_{t_i}^{t_f} \mathscr{L}(x^j(t), \dot{x}^j(t), t) dt \tag{5.47}$$

is the time integral of the classical Lagrangian of the system calculated along the path $x^j(t)$ between the fixed times t_i and t_f. Notice that the action appears in the exponent of eq. 5.46 in units of $\hbar$.

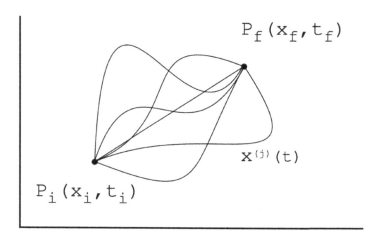

Figure 5.7   Few paths to go from P_i to P_f.

We see immediately from eq. 5.21, that S_{fi} is the action or, equivalently, Hamilton's principal function of the mechanical system.

Eq. 5.47 is a recipe for calculating the action associated to each path: in other words, given a path x^j, eq. 5.47 associates a (real) number to it. The action 5.47 is therefore a functional. We can now finally express the amplitude for a system to start at point (x_i, t_i) to end at point (x_f, t_f):

$$K(x_i, t_i, x_f, t_f) = \sum_j F_{fi}^j e^{\frac{i}{\hbar} S_{fi}^j} \tag{5.48}$$

The function $K(x_i, t_i, x_f, t_f)$ is called spacetime *propagator* because describes the "propagation" of the system from the initial spacetime location to the final spacetime location[7].

In fig. 5.7 only a few possible paths are shown. In theory, any path will contribute to the calculation of the probability amplitude, and therefore to the actual path followed by the particle. The theory shows that paths for which the propagation happens at a speed faster than the speed of light are associated with nonzero propagator therefore implying superluminal motion or superluminal transmission of information clearly violating special relativity. In practice, particles, associated with each path are called *virtual particles*, and it can be shown that special relativity is not violated because virtual particles do not carry information.

Let us now go back to our search for the equation governing wavefunctions and *define* the wavefunction as:

$$\psi(x, t) \equiv \frac{K}{F} = e^{\frac{i}{\hbar} S(x, t)} \tag{5.49}$$

---

[7]In more advanced quantum mechanics, which is not discussed here, the propagator is used in the so called *Feynman diagrams* in quantum field theory.

We now show that this wavefunction allows us to make a direct connection between the HJ equation and the new wave equation we are after. Let's report here, for convenience, the HJ equation 5.32 discussed earlier for just the x coordinate:

$$\frac{1}{2m}\left(\frac{\partial S}{\partial x}\right)^2 + V(x) + \frac{\partial S}{\partial t} = 0 \tag{5.50}$$

We notice that eq. 5.49 expresses a relationship between the action S and the wavefunction ψ. We start by taking the log of both sides of eq. 5.49:

$$S = -i\hbar \ln \psi \tag{5.51}$$

Let's restrict ourselves to just one space coordinate x for simplicity and let's take the partial derivative with respect to x and t of eq. 5.51:

$$\begin{aligned}\frac{\partial S}{\partial x} &= -i\hbar \frac{1}{\psi}\frac{\partial \psi}{\partial x} \\ \frac{\partial S}{\partial t} &= -i\hbar \frac{1}{\psi}\frac{\partial \psi}{\partial t}\end{aligned} \tag{5.52}$$

Let's invert the first equation in 5.52 and differentiate a second time with respect to x:

$$\begin{aligned}\frac{\partial \psi}{\partial x} &= \frac{i}{\hbar}\psi\frac{\partial S}{\partial x} \\ \frac{\partial^2 \psi}{\partial x^2} &= -\frac{\psi}{\hbar^2}\left(\frac{\partial S}{\partial x}\right)^2\end{aligned} \tag{5.53}$$

the missing term $\frac{i}{\hbar}\psi\frac{\partial^2 S}{\partial x^2}$ is equal to zero because $\frac{\partial^2 S}{\partial x^2} = \frac{\partial p_x}{\partial x} = m\frac{\partial}{\partial x}\left(m\frac{dx}{dt}\right) = m\frac{d}{dt}\left(\frac{\partial x}{\partial x}\right) = 0$. We now insert the second equation of 5.53 and the second equation of 5.52 into the HJ equation 5.29 to have:

$$i\hbar\frac{\partial \psi}{\partial t} = -\frac{\hbar^2}{2m}\frac{\partial^2 \psi}{\partial x^2} + V\psi \tag{5.54}$$

which is the **Schrödinger equation** for the wavefunction ψ representing the amplitude associated to a Feynman's path. If we were to consider the 3 spatial dimensions, then Schrödinger equation is:

$$i\hbar\frac{\partial \psi}{\partial t} = -\frac{\hbar^2}{2m}\nabla^2\psi + V\psi \tag{5.55}$$

where:

$$\nabla^2 = \frac{\partial}{\partial x^2} + \frac{\partial}{\partial y^2} + \frac{\partial}{\partial z^2}. \tag{5.56}$$

Schrödinger equation 5.55 is a linear partial differential equation for the wavefunction $\psi(x, y, z, t)$.

In classical mechanics, the dynamics of a system made of N particles of mass m_i, with $i = 1, ..., N$, is governed by Newton's laws and the state of a system is defined by the knowledge of all positions and momenta (r_i, p_i). In quantum mechanics, the state of a system is known when the wavefunction of the system is given. The time evolution of the wavefunction is described by Schrödinger equation. The wavefunction is a probability amplitude and, in general, is a complex function[8]. The square of the modulus of the wavefunction $|\psi(r,t)|^2$ is the probability density. This means that $|\psi(r,t)|^2 dr$ is the probability of finding the system between the positions r and $r + dr$ at the time t. In order to be correctly representing a probability, we must have the normalization condition:

$$\int |\psi(r,t)|^2 \, dr = 1. \tag{5.57}$$

Since we have already mentioned that the wavefunction in general must be complex, then eq. 5.57 can be written as:

$$\int \psi^*(r,t)\psi(r,t) \, dr = 1 \tag{5.58}$$

where $\psi^8(r,t)$ is the complex conjugate of $\psi(r,t)$.

The linearity of Schrödinger equation implies that if ψ_1 and ψ_2 are two solutions, then the sum $\psi = \psi_1 + \psi_2$ is also a solution. This characteristic has many important consequences that will be explored later.

5.2.4 PLANE WAVES

In this section we show how Schrödinger equation is related to the classical energy and the important role of plane wave solutions. We have seen that, for a nonrelativistic particle of mass m, its total energy in one dimension x is the sum of the kinetic energy T and the potential energy V:

$$E = T + V = \frac{1}{2}mv^2 + V = \frac{p^2}{2m} + V. \tag{5.59}$$

If we now insert in eq. 5.59 the de Broglie relation $p = \hbar k$, we have:

$$E = \frac{p^2}{2m} + V = \frac{\hbar^2 k^2}{2m} + V \tag{5.60}$$

and using Planck formula, eq. 5.60 becomes:

$$E = h\nu = \hbar\omega = \frac{\hbar^2 k^2}{2m} + V. \tag{5.61}$$

Let's assume that a plane wave of the form:

$$\psi(x,t) = Ae^{i(kx - \omega t)} \tag{5.62}$$

---

[8]We will briefly discuss at the end of the book why the wavefunction has to be a complex function.

where the amplitude A needs to be normalized to make sure that square modulus represents a probability. Let us study the time derivative of the plane wave:

$$\frac{\partial}{\partial t}\psi(x,t) = \frac{\partial}{\partial t}[Ae^{i(kx-\omega t)}] = -i\omega\psi(x,t). \tag{5.63}$$

Eq. 5.63 can be inverted to express the de Broglie angular frequency:

$$\omega = \frac{i}{\psi}\frac{\partial\psi}{\partial t} \tag{5.64}$$

where, for simplicity, we are not writing explicitly the dependency of ψ from (x,t).

The space derivatives are:

$$\frac{\partial\psi}{\partial x} = ik\psi$$
$$\frac{\partial^2\psi}{\partial x^2} = -k^2\psi \tag{5.65}$$

from which we extract k^2:

$$k^2 = -\frac{1}{\psi}\frac{\partial^2\psi}{\partial x^2}. \tag{5.66}$$

Substituting eq. 5.64 and eq. 5.66 into eq. 5.61 we have:

$$\frac{i\hbar}{\psi}\frac{\partial\psi}{\partial t} = \frac{\hbar^2}{2m}\left[-\frac{1}{\psi}\frac{\partial^2\psi}{\partial x^2}\right] + V\psi \tag{5.67}$$

and, after eliminating the common ψ from both sides of eq. 5.67 we have:

$$i\hbar\frac{\partial\psi}{\partial t} = -\frac{\hbar^2}{2m}\left[\frac{\partial^2\psi}{\partial x^2}\right] + V\psi \tag{5.68}$$

which is exactly equal to Schrödinger equation 5.54.

As an example, let's see what is the wavefunction of a free particle of mass m and momentum p. It should be evident, by now, that experiments are showing that some sort of wave, with de Broglie wavelength, is associated with any massive particle. Assuming that our free particle is moving along the $x-$direction, with momentum $p = \hbar k$ where $k = \frac{2\pi}{\lambda}$, the simplest wavefunction would be a plane wave of the form:

$$\psi(x,t) = Ae^{i(kx-\omega t)} \tag{5.69}$$

Notice that this wavefunction is complex. In fact, using Euler's formula[9] eq. 5.69 can be written as:

$$\psi(x,t) = A(\cos(kx-\omega t) + i\sin(kx-\omega t)) \tag{5.70}$$

---

[9]Euler's formula is $e^{i\alpha} = \cos\alpha + i\sin\alpha$.

Notice also that a wavefunction of the form 5.69 guarantees that the probability $|\psi|^2 = \psi^*\psi = |A|^2 = $ constant, where ψ^* is the complex conjugate[10] of ψ, is a constant value corresponding to the fact that the particle must be somewhere between $-\infty$ and $+\infty$ with equal probability. We go ahead of us for a moment and we notice that if the plane wave 5.69 represents a particle of exactly defined wavenumber k, then we cannot tell where the particle is located in space.

There is a convincing argument to show that a wavefunction of the form 5.69 (plane wave) is needed to represent a free particle propagating in the direction of positive x. In fact, let us study a simpler wavefunction of the form:

$$\psi(x,t) = A_1 \cos(kx - \omega t + \delta_1) \tag{5.71}$$

where δ_1 is a phase. If we wanted to study the propagation in the direction of the negative x then the wavefunction would be:

$$\psi(x,t) = A_2 \cos(kx + \omega t + \delta_2). \tag{5.72}$$

We now state a fundamental assumption about the wavefunction: **the knowledge of the wavefunction at the time $t = 0$ represents a complete knowledge of our system**, including its time evolution. In other words, knowing $\psi(x,0)$ is enough to predict $\psi(x,t)$. If this assumption is valid, then we cannot represent plane waves with real expressions of the form 5.71 or 5.72. The main problem[11] consists in recognizing that 5.71 represents a particle moving in the positive x direction while 5.72 represents a particle moving in the opposite direction. We now require, as it seems reasonable, that these two functions are linearly independent for any t, i.e. they represent distinct particles traveling in opposite directions. It follows that the requirement of linear independence is violated at the special time $t = \frac{\alpha_1 - \alpha_2}{2\omega}$ when the two functions are proportional and so are linearly dependent.

The simplest choice to avoid this problem would be to add a sinusoidal term with the same argument $(kx - \omega t)$, i.e. the same de Broglie wavelength moving at the same speed, to both expressions:

$$\begin{aligned} \psi_1(x,t) &= A_1[\cos(kx - \omega t) + \delta_1 \sin(kx - \omega t)] \\ \psi_2(x,t) &= A_2[\cos(kx - \omega t) + \delta_2 \sin(kx - \omega t)] \end{aligned} \tag{5.73}$$

and we now require that the two expression in 5.73 are linearly independent at all times. This translates into finding the two constants δ_1 and δ_2 such that a plane wave $\psi_1(x,t)$ represents a particle propagating along the positive x direction without any admixture of $\psi_2(x,t)$ propagating in the opposite direction. This means also that the plane wave representing the particle at time t, $\psi(x)$ is just a multiple of the plane wave at $t = 0$, i.e. $\psi(x,t) = a\psi(x,0)$. More in general, if we indicate with $\varepsilon = -\omega t$ the time dependence term, we must have, for all x and ε:

$$\cos(kx + \varepsilon) + \delta_1 \sin(kx + \varepsilon) = a_1(\varepsilon)(\cos kx + \delta_1 \sin kx). \tag{5.74}$$

---

[10]The complex conjugate of a complex number $h = a + ib$ is $h^* = a - ib$.

[11]For an extended treatment of this argument see the excellent book by Merzbacher [28].

Using $\cos(a \pm b) = \cos a \cos b \mp \sin a \sin b$ and $\sin(a \pm b) = \sin a \cos b \pm \cos a \sin b$, eq. 5.74 becomes:

$$(\cos \varepsilon + \delta_1 \sin \varepsilon) \cos kx + \sin \varepsilon (-\sin kx + \delta_1 \cos \varepsilon) = a_1(\varepsilon) \cos kx + a_1(\varepsilon) \delta_1 \sin kx.$$
$$(5.75)$$

To be valid for all x and ε, the coefficients of $\cos kx$ and $\sin kx$ must be equal:

$$\cos \varepsilon + \delta_1 \sin \varepsilon = a_1(\varepsilon)$$
$$\delta_1 \cos \varepsilon - \sin \varepsilon = a_1(\varepsilon) \delta_1$$
$$(5.76)$$

and it is easy to verify that the condition 5.76 is satisfied only if $\delta_1 = \pm i$. Analog result is obtained if we impose a similar condition for $\psi_2(x,t)$. Choosing $\delta_1 = +i$ and $\delta_2 = -i$ for, respectively, plane waves propagating toward positive and negative x, we justify the choice for plane waves:

$$\psi_1(x,t) = A_1 e^{i(kx - \omega t)}$$
$$\psi_2(x,t) = A_2 e^{-i(kx - \omega t)}.$$
$$(5.77)$$

5.3 HEISENBERG UNCERTAINTY PRINCIPLE

We have seen in the previous chapters that there is overwhelming experimental evidence that particles are associated with waves in order to explain a variety of experiments where small particles, like electrons, experience interference phenomena. Scattering of electrons from crystal lattices or double slit interference are two of many examples. In analogy of e.m. diffraction from a small aperture (see fig. 5.4), diffraction of particles from a single slit (or a small hole) is also an experimental fact. In fig. 5.8 we have a particle of known horizontal momentum p going through a screen where there is an aperture of dimension $D \sim \lambda_p$, where λ_p is the associated de Broglie wavelength. Before passing through the aperture, the momentum vector is $p = (p_x, 0)$, i.e. the momentum has only the horizontal component along the propagation axis x.

As we have stressed previously, once the particle goes through the aperture D there is no way to accurately predict where the particle will land on the observing screen. The best we can do is to send many particles with the same momentum and record the position of where the particle land on the observing screen. When enough particles have been recorded a typical diffraction pattern will emerge, where the maximum number of particles is recorded at $y = 0$, i.e. the particle have crossed the aperture with no deviation, as expected classically. However, and in contrast with classical expectations, there will be particles scattered even at large angles.

In analogy with the diffraction of e.m. waves, the first zero in the diffraction pattern happens when the spreading angle $\Delta\theta$ is of the order of λ/D.

Following Feynman [16], let us now study this particle diffraction experiment and obtain an order of magnitude expression for our knowledge of the system in order to make future predictions of the dynamics of the particle. We already have guessed, by looking at the right panel of fig. 5.8, that the more we try to localize the particle as it

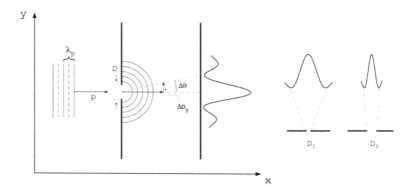

Figure 5.8 Diffraction of particles with momentum p from a small aperture $D \sim \lambda$. The right panel shows the inverse relationship between aperture and spreading angle when $D_1 <$ D_2.

goes through the aperture, i.e. the smaller we make D, the larger will be the spread of momentum due to diffraction of the wave associated with the particle. It is natural to make the hypothesis that the product of the knowledge of the position times the knowledge of the momentum is constant so that there is an inverse proportionality. The remarkable fact that the constant is Planck's constant should not surprise us by now.

When the particle is approaching the aperture, we know its momentum p which is totally in the x direction. At the time of passing through the aperture, our knowledge of the position of the particle is $\Delta y \approx D$. After the particle has passed through the aperture, we know that the momentum is not anymore just along the x direction but it is subject to a spread that we can roughly estimate as equal to $p_y = p\Delta\theta = p\frac{\lambda}{D}$ along the y direction. But we know from de Broglie that the product $p\lambda = h$ and so we have the interesting result that the product of the uncertainties in the position and momentum is:

$$\Delta y \, \Delta p_y \approx h. \tag{5.78}$$

This is the so called *Heisenberg uncertainty principle* (HUP) and was first proposed by Heisenberg[12] in the year 1927 [20]. In order to be consistent with experiments, the exact Heisenberg uncertainty relationship is an inequality:

$$\Delta y \, \Delta p_y \geq \frac{\hbar}{2}. \tag{5.79}$$

More in general, HUP tells us that there is a fundamental limit to how well we know the value of a pair of two measurable physical quantities given the initial conditions. Notice that, in the case of position and momentum, the HUP tells us that the

---

[12]Werner K. Heisenberg (1901-1976) was a German theoretical physicist and one of the fathers of quantum mechanics. In addition to the uncertainty principle, Heisenberg formulated a theory of quantum mechanics based on matrix algebra that will be discussed later in this book.

inequality 5.79 applies to the projection of the momentum along the direction of the position variable. If we were to consider the x coordinate, for example, then the HUP is written as $\Delta x \, \Delta p_x \geq \frac{\hbar}{2}$, and similar expression for the z coordinate.

A more accurate statement of the HUP must include a better definition of the uncertainty symbol Δ. In a statistical way, if we introduce the standard deviation obtained after many experimental measurements of, for example, the position of our particle, we have:

$$\sigma_x = \sqrt{\frac{1}{N}\sum_1^N (x_i - \bar{x})^2} \tag{5.80}$$

in case of a discrete set and where $\bar{x}$ is the average value of x. If we have a continuous variable, then the standard deviation is:

$$\sigma_x = \sqrt{\int (x - \bar{x})^2 g(x) dx} \tag{5.81}$$

where $g(x)$ is the probability density function, i.e. $g(x)dx$ is the probability of finding the particle between x and $x + dx$, and $\bar{x} = \int x g(x) dx$. Using the definition 5.80 or 5.81 we have that a better definition of the HUP[13] is:

$$\sigma_x \, \sigma_{p_x} \geq \frac{\hbar}{2}. \tag{5.82}$$

We can write more generally eq. 5.79 for the pair of variables (q, p) of Hamilton's formalism:

$$\sigma_q \, \sigma_p \geq \frac{\hbar}{2}. \tag{5.83}$$

The variables (q, p) are *conjugate variables* in the sense that they are connected by a transformation. In fact, the first eq. 5.24 tells us that the momentum is the partial derivative of the action S with respect to the generalized coordinate q, re-introducing interestingly the action S.

Are there other pairs of conjugate variables for which an HUP can be written? The answer is yes and the second equation in 5.24 gives us:

$$\sigma_E \, \sigma_t \geq \frac{\hbar}{2}. \tag{5.84}$$

This last HUP can be interpreted in many ways. It means, for example, that the determination of the energy E with an accuracy σ_E requires *at least* a time $\sim \hbar/\sigma_E$. A more subtle interpretation consists in requiring that energy conservation is violated providing that the violation is shorter that the time $\hbar/\sigma_E$, etc. Obviously all these effects are important only at atomic scales given the small value of $\hbar$.

Let us mention another conjugate pair connected through the action S. In the first equation 5.13, we have seen a relationship between generalized momentum p and

---

[13]This is also known as Kennard inequality.

the derivative of the action with respect to the generalized coordinate q. We introduce a new couple of canonical conjugate variables (L, θ) obtained through a canonical transformation $(p, q) \rightarrow (L, \theta)$, where L is the angular momentum and θ is the angle. It can be shown that $L = \frac{\partial S}{\partial \theta}$ and so, in analogy with the momentum/position and energy/time, we have an additional HUP:

$$\sigma_L \, \sigma_\theta \geq \frac{\hbar}{2}. \tag{5.85}$$

Eq. 5.85 tells us that an accurate determination of the angular momentum of a particle in orbit involves a corresponding loss of the knowledge of its angular position.

Finally, we notice that the conjugate variables we just described, i.e. for example (x, p_x), (E, t) and (L, θ) are connected through the Noether theorem discussed in chapter 1. This reinforces, once more, the fundamental role of the action S as a bridge between classical and quantum mechanics through conjugate variables, the HUP and the symmetries of Noether theorem .

5.4 OPERATORS AND EXPECTATION VALUE

Let's rewrite Schrödinger eq. 5.68 in a slightly different way:

$$i\hbar \left[\frac{\partial}{\partial t} \right] \psi = -\frac{\hbar^2}{2m} \left[\frac{\partial^2}{\partial x^2} + V \right] \psi \tag{5.86}$$

and compare with eq. 5.60. In the transition from classical mechanics to quantum mechanics it seems natural to make the following associations:

$$E \rightarrow i\hbar \frac{\partial}{\partial t} \psi$$
$$T + V \rightarrow -\frac{\hbar^2}{2m} \left[\frac{\partial^2}{\partial x^2} + V \right] \psi. \tag{5.87}$$

More specifically, the energy seems to be associated with the partial derivative with respect to time of the wavefunction ψ while the classical scalar function V, in quantum mechanics, becomes $-\frac{\hbar^2}{2m} V \psi$ and the classical kinetic energy $\frac{p^2}{2m}$ becomes $-\frac{\hbar^2}{2m} \frac{\partial^2}{\partial x^2} \psi$. This last quantity seems to associate the momentum with the partial derivative with respect to the coordinate x, i.e. $p \rightarrow -i\hbar \frac{\partial \psi}{\partial x}$.

More in general, in 3 dimensions, it is customary to make the following associations:

$$\hat{E} \rightarrow i\hbar \frac{\partial}{\partial t}$$
$$\hat{p} \rightarrow -i\hbar \frac{\partial}{\partial r} = -i\hbar \nabla \tag{5.88}$$
$$\hat{r} \rightarrow r$$

where we indicated with an hat above the symbol the fact that they are quantum mechanical objects supposed to be applied *to their right* to the wavefunction ψ. We have therefore introduced the concept of *operators*, i.e. mathematical objects that, when applied to a function produce another function. They are an extension of the more familiar concept of function $y = f(x)$ where the symbol $f(x)$ means that we associate a number y to a number x according to the recipe $f(x)$. In the case of operators, we associate a function $\phi(x,t) = \hat{A}\psi(x,t)$, meaning that the operator $\hat{A}$ transforms the function $\psi(x,t)$ into a new function $\phi(x,t)$.

It is important to state that operators act on the function that is written on their right. Classical measurable quantities (observables) like the total energy, the momentum, the position, etc., become operators in quantum mechanics. More in general, operators are mathematical quantities that act on the wavefunction and produce another wavefunction. Notice that the position operator $\hat{r}$ is just the multiplication of r by the wavefunction, i.e. $\hat{r}\psi = r \cdot \psi$.

We can push a bit further the probabilistic interpretation of the wavefunction by calculating the so called *expectation values* which is defined as the generalization of the weighted average. For a continuous variable, like the position x, the expectation value is usually indicated with $\langle x \rangle$ and is:

$$\langle \hat{x} \rangle = \int \hat{x}\, |\psi(r,t)|^2 dx = \int \psi^*(x,t)\, \hat{x}\, \psi(x,t) dx \tag{5.89}$$

which means either the probability of the result of a single experiment or the average of many experiments done on many independent identical systems. We have that the expectation values of the energy and the momentum are:

$$\langle E \rangle = i\hbar \int \psi^* \frac{\partial \psi}{\partial t}\, dr$$
$$\langle p \rangle = -i\hbar \int \psi^* \nabla \psi\, dr. \tag{5.90}$$

The Schrödinger equation is equivalent to the classical expression of the total energy, $E = T + V$ with the substitution represented in eq. 5.88.

Let us now introduce two operators: kinetic energy $\hat{T}$ and potential energy $\hat{V}$. If the potential energy $V = V(r)$ is a function of only the position, then its quantum mechanical operator is just a multiplication and therefore:

$$\hat{V} = V(r). \tag{5.91}$$

The kinetic energy operator can be constructed by using the momentum operator. Starting with $T = \frac{p^2}{2m}$ we consider the square of the operator $\hat{p}$ as its application twice in sequence, i.e. $\hat{p}^2\psi = \hat{p}(\hat{p}\psi)$. Therefore the kinetic energy operator can be written as:

$$\hat{T} = \frac{\hat{p}^2}{2m} = -i\hbar\nabla \cdot (-i\hbar\nabla)\frac{1}{2m} = -\frac{\hbar^2}{2m}\nabla^2 \tag{5.92}$$

We can finally express a very important operator, the Hamiltonian $\hat{H}$, as:

$$\hat{H} = \hat{T} + \hat{V} = -\frac{\hbar^2}{2m}\nabla^2 + V(r). \tag{5.93}$$

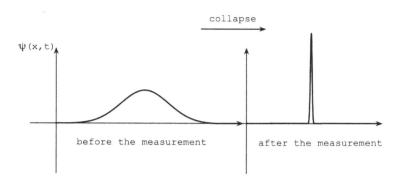

Figure 5.9   Wavefunction collapse due to the act of measurements.

This will allow us to express Schrödinger equation in a very compact form:

$$i\hbar \frac{\partial \psi}{\partial t} = \hat{H}\psi. \tag{5.94}$$

5.4.1   MEASUREMENTS AND COMMUTATORS

If we make the assumption that the wavefunction $\psi(r,t)$ represents the status of a physical system at the time t, we have seen that an operator representing some measurable physical quantity acts on its right on the wavefunction. One of the basic postulates of quantum mechanics consists in assuming that an operator acting on the wavefunction is equivalent to make a measurement of the physical quantity associated with the operator at the time t. For example, a measurement of the position will be achieved by applying the operator $\hat{r}$ to the wavefunction: $\hat{r}\psi(r,t)$. As another example, a measurement of the momentum of a particle is indicated as $\hat{p}\psi(r,t) = -i\hbar\nabla\psi(r,t)$.

We might wonder what happens to the wavefunction once we have finished our measurements. We will not treat the interesting and vast subject of the various interpretations of quantum mechanics but we limit ourselves to the very popular *Copenhagen interpretation* which states that while the wavefunction before the measurements represents the probability amplitude of finding the system in a particular state, after the measurements the wavefunction has changed dramatically (see fig. 5.9). Before the measurement the system is in a state for which we can only know the probability of achieving a certain value for the measured quantity. After the measurements, we now know that the system is actually in one particular state and the wavefunction must therefore change to represent the new state. In the case of fig. 5.9 if the wavefunction on the left represents, for example, the position of a particle, after the measurement we know that the particle is in a particular position and this is represented by a very peaked wavefunction around the measured position.

Working with operators has an interesting consequence: if we perform two different measurements on a system, the end result depends on the order in which we make the measurements. For example, if we first measure the momentum and then the position of a particle, the result will be different if we first measure the position and then the momentum. This is expressed mathematically by:

$$\hat{x}\hat{p}\psi \neq \hat{p}\hat{x}\psi \tag{5.95}$$

which is read like this: we first measure the momentum $\hat{p}\psi$ and then the position $\hat{x}(\hat{p}\psi)$. The result is different if we first measure the position $\hat{x}\psi$ and then measure the momentum $\hat{p}(\hat{x}\psi)$. Therefore the order by which we apply operators on the wavefunction is very important. The *commutator* of two operators tells us what is this difference and is indicated by the symbol $[\hat{x}, \hat{p}]$:

$$[\hat{x}, \hat{p}]\psi = \hat{x}\hat{p}\psi - \hat{p}\hat{x}\psi. \tag{5.96}$$

We will see later in the book that the commutator between operators plays a fundamental role in quantum mechanics. For this reason, we give here a list of properties. Given three operators A, B, C and a constant c, we have that:

$$\begin{aligned}
[A, c] &= 0 \\
[A, A] &= 0 \\
[A, B] &= -[B, A] \\
[cA, B] &= [A, cB] = c[A, B] \\
[A, B \pm C] &= [A, B] \pm [A, C] \\
[AB, C] &= A[B, C] + [A, C]B \\
[A, BC] &= B[A, C] + [A, B]C
\end{aligned} \tag{5.97}$$

There is a relationship between the commutators and the HUP. In order to show this, let's simplify our notation and write the expectation value for an operator $\langle \hat{A} \rangle = \int \psi^* \hat{A} \psi$ in a simplified way as $\langle \hat{A} \rangle = \psi^* \hat{A} \psi$ where we do not explicitly write the sign of integral. Let's now write the variance (i.e. the square of the standard deviation) of the operator $\hat{x}$ and the operator $\hat{p}_x$ from eq. 5.89:

$$\begin{aligned}
\sigma_x^2 &= \psi^* (\hat{x} - \langle \hat{x} \rangle)^2 \psi \\
\sigma_{p_x}^2 &= \psi^* (\hat{p}_x - \langle \hat{p}_x \rangle)^2 \psi
\end{aligned} \tag{5.98}$$

Let us introduce two operators $\hat{H}$ and $\hat{K}$:

$$\begin{aligned}
\hat{H} &= \hat{x} - \psi^* \hat{x} \psi \\
\hat{K} &= \hat{p}_x - \psi^* \hat{p}_x \psi.
\end{aligned} \tag{5.99}$$

It is easy to show that eq. 5.98 becomes:

$$\begin{aligned}
\sigma_x^2 &= \psi^* (\hat{x} - \langle \hat{x} \rangle)^2 \psi = \psi^* \hat{H}^2 \psi \\
\sigma_{p_x}^2 &= \psi^* (\hat{p}_x - \langle \hat{p}_x \rangle)^2 = \psi^* \hat{K}^2 \psi.
\end{aligned} \tag{5.100}$$

Let us build another operator $\hat{Q} = \hat{H} + i\alpha\hat{K}$ and study its square modulus which must be positive, i.e. $\psi^*Q^* \cdot Q\psi \geq 0$. Notice that we have written the complex conjugate operator $\hat{Q}^*$ on the right of the wavefunction (also complex conjugated) ψ^* with the convention that complex conjugate operators act on the complex conjugate wavefunctions on their left. We have:

$$\psi^*Q^* \cdot Q\psi = \psi^*(\hat{H} - i\alpha\hat{K})(\hat{H} + i\alpha\hat{K})\psi$$
$$= \psi^*\hat{H}^2\psi + \alpha^2\psi^*\hat{K}^2\psi - i\alpha\psi^*\hat{K}\hat{H}\psi + i\alpha\psi^*\hat{H}\hat{K}\psi \qquad (5.101)$$
$$= \sigma_x^2 + \alpha^2\sigma_{p_x}^2 + i\alpha\psi^*[H,K]\psi \geq 0.$$

It is easy to verify that $[H,K] = [\sigma_x, \sigma_{p_x}]$ and so eq. 5.101 becomes:

$$\psi^*Q^* \cdot Q\psi = \sigma_x^2 + \alpha^2\sigma_{p_x}^2 + i\alpha\psi^*[H,K]\psi \geq 0. \qquad (5.102)$$

We now ask what is the value of α that minimizes the inequality 5.102? This is calculated by setting to zero the derivative with respect to α of the expression in eq. 5.102:

$$\frac{\partial}{\partial\alpha}(\psi^*Q^* \cdot Q\psi) = \frac{\partial}{\partial\alpha}(\sigma_x^2 + \alpha^2\sigma_{p_x}^2 + i\alpha\psi^*[H,K]\psi) = 0. \qquad (5.103)$$

Taking the derivative we have:

$$2\alpha\sigma_{p_x}^2 + i\psi^*[H,K]\psi) = 0 \qquad (5.104)$$

From which we obtain $\alpha = -\frac{i}{2\sigma_{p_x}^2}\psi^*[H,K]\psi$ to be plugged back into eq. 5.102 to give, after a bit of algebra, a very important relation:

$$\sigma_x\sigma_{p_x} \geq \frac{i}{2}[\hat{x}, \hat{p}_x]. \qquad (5.105)$$

Eq. 5.105 has been written for the two quantum operators $\hat{X}$ and $\hat{p}_x$ but it is a more general result. Given two noncommuting operators $\hat{A}$ and $\hat{B}$, i.e. for which $[\hat{A}, \hat{B}] \neq 0$, then the commutator is:

$$\sigma_A\sigma_B \geq \frac{i}{2}[\hat{A}, \hat{B}]. \qquad (5.106)$$

The HUP is recovered if the commutator obeys:

$$[\hat{x}, \hat{p}] = i\hbar \qquad (5.107)$$

as it can immediately be verified by substitution.

Let's list a few properties of commutators that will be useful later. We have that:

I Any operator $\hat{A}$ commutes with a scalar, $[\hat{A}, c] = 0$
II Given three operators $\hat{A}, \hat{B}, \hat{C}$, $[\hat{A}, \hat{B}\hat{C}] = [\hat{A}, \hat{B}]\hat{C} + \hat{B}[\hat{A}, \hat{C}]$
III Any operator $\hat{A}$ commute with itself, $[\hat{A}, \hat{A}] = 0$.

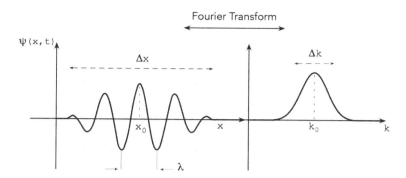

Figure 5.10 Left: The wavefunction (wavepacket) representing a particle localized in a region Δx with associated de Broglie wavelength λ. Right: The Fourier transform of the wavepacket.

5.5 WAVEPACKETS AND EHRENFEST'S THEOREM

So far we have introduced the wavefunction as a complex amplitude whose square modulus, if properly normalized, gives the probability of finding the particle in a particular location. The probability, on the other hand, subtends the concept that we are dealing with experiments repeated many times under the exact same initial conditions. Therefore we might wonder what is the meaning of the wavefunction of a single particle? One idea would be to identify a particle with a sort of localized bunch of waves, as shown in fig. 5.10. In this representation, the particle would be identified by a wavepacket of spatial spread Δx with a de Broglie wavelength λ. Its Fourier transform represents the harmonic content needed, i.e. how many sine and cosine waves with appropriate phases need to be added together to reproduce the wavepacket. The basic idea would be that outside the region Δx, the superposition of the various waves produces a destructive interference while within the region Δx the interference is constructive.

The spread in position Δx is inversely proportional to the spread in momentum $\Delta p = \hbar \Delta k$: in fact, the more we try to localize the particle by making the spread Δx smaller, the larger will be the spread in momentum because we need more and more waves to interfere destructively outside Δx, and vice versa. The limiting case would be a delta function to localize exactly the particle at the cost of complete loss of knowledge of its momentum and vice versa.

We should not be surprised about this result because it is related to a general properties of the Fourier transform, namely the so called *bandwidth theorem* which states that *the product of the width σ_x of a function times the widths σ_p of its Fourier transform obeys the inequality $\sigma_x \sigma_p \geq \frac{1}{2}$*. The Fourier transform (FT) of a function $f(x)$ is defined as:

$$\phi(p) = \int_{-\infty}^{+\infty} f(x) e^{-i2\pi px} dx \qquad (5.108)$$

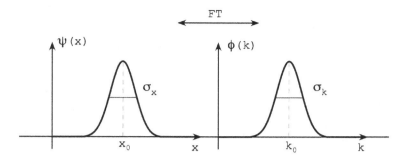

Figure 5.11   Gaussian wavepacket and its Fourier transform.

and its inverse is defined as:

$$f(x) = \int_{-\infty}^{+\infty} \phi(p)e^{i2\pi px}dx \tag{5.109}$$

and the widths are given by eq. 5.81 where $g(x) = |f(x)|^2$. In terms of wavenumber k and if we want to have functions and their FT to have the same normalization (symmetric form) we have:

$$f(x) = \frac{1}{\sqrt{2\pi}} \int_{-\infty}^{+\infty} F(k)e^{ikx}dk$$
$$F(k) = \frac{1}{\sqrt{2\pi}} \int_{-\infty}^{+\infty} f(x)e^{-ikx}dx. \tag{5.110}$$

With the definition 5.110 the normalization is such that:

$$\int_{-\infty}^{+\infty} f^*(x)f(x)dx = \int_{-\infty}^{+\infty} F^*(k)F(k)dk. \tag{5.111}$$

A generic wavepacket is described by the function $f(x)$ whose harmonic content is defined by the function $F(k)$. The symmetry between the two functions $f(x)$ and $F(k)$ is such that if the probability for finding the particle between x and $x + dx$ is $f^*(x)f(x)dx = |f(x)|^2 dx$ then the probability for the particle to have wavenumber between k and $k + dk$ is $F^*(k)F(k)dk = |F(k)|^2 dk$.

Let's show that for a Gaussian wavepacket $\psi(x)$ and its FT $\phi(k)$ as shown in fig. 5.11, we have $\sigma_x \sigma_k = \frac{1}{2}$ corresponding to the minimum uncertainty product. We are assuming that the spread in x−coordinate is the standard deviation σ_x while the spread in k−coordinate is the standard deviation σ_k.

A Gaussian wavepacket in k space can be written as:

$$\phi(k) = \left(\frac{2a}{\pi}\right)^{\frac{1}{4}} e^{-a(k-k_0)^2} \tag{5.112}$$

where a is a constant. The corresponding wavepacket in x space is obtained by FT the function 5.112:

$$\psi(x) = \frac{1}{\sqrt{2\pi}} \left(\frac{2a}{\pi}\right)^{\frac{1}{4}} \int_{-\infty}^{+\infty} e^{-a(k-k_0)^2} e^{ikx} dk. \tag{5.113}$$

The probability associated with the wavepacket 5.112 is:

$$|\phi(x)|^2 = \phi^*(x)\phi(x) = \sqrt{\frac{2a}{\pi}} e^{-2a(k-k_0)^2} \tag{5.114}$$

to be compared with a normalized Gaussian distribution in $k-$coordinate:

$$G(k) = \frac{1}{\sqrt{2\pi\sigma_k^2}} e^{-\frac{(k-k_0)^2}{2\sigma_k^2}} \tag{5.115}$$

The standard deviation σ_k can be inferred by comparing eq. 5.115 with eq. 5.114. We have immediately that $\sigma_k = \frac{1}{2\sqrt{a}}$.

Working out σ_x is not as immediate because of the definition of $\psi(x)$ in eq. 5.113 which contains an integral and a complex exponential e^{ikx}.

Let's write the well-known integral of a Gaussian distribution in $k-$coordinate:

$$\int_{-\infty}^{+\infty} e^{-ak^2} dk = \sqrt{\frac{\pi}{a}}. \tag{5.116}$$

Our task is to change variables in eq. 5.113 in such a way that the integral can be calculated using eq. 5.116. Let's make the following change of variable:

$$k' = k - k_0 - \frac{ix}{2a} \tag{5.117}$$

with $dk' = dk$. The exponent of the integral in eq. 5.113 becomes:

$$
\begin{aligned}
-a(k - k_0)^2 + ikx &= -a(k' + \frac{ix}{2a})^2 + ikx \\
&= -a(k'^2 + \frac{ix}{a}k' - \frac{x^2}{4a^2}) + ikx \\
&= -ak'^2 - ik'x + \frac{x^2}{4a^2} + ikx \\
&= -ak'^2 + ix(k - k') + \frac{x^2}{4a^2} \\
&= -ak'^2 + ix(k_0 + \frac{ix}{2a}) + \frac{x^2}{4a^2} \\
&= -ak'^2 + ik_0x - \frac{x^2}{2a}
\end{aligned}
\tag{5.118}
$$

and we can rewrite eq. 5.113 as:

$$\psi(x) = \left(\frac{a}{2\pi^3}\right)^{\frac{1}{4}} e^{ik_0x} e^{-\frac{x^2}{4a}} \int_{-\infty}^{+\infty} e^{-ak'^2} dk. \tag{5.119}$$

Using eq. 5.116, eq. 5.119 becomes:

$$\psi(x) = \left(\frac{a}{2\pi^3}\right)^{\frac{1}{4}} e^{ik_0x} e^{-\frac{x^2}{4a}} \sqrt{\frac{\pi}{a}} = \left(\frac{1}{2\pi a}\right)^{\frac{1}{4}} e^{ik_0x} e^{-\frac{x^2}{4a}}. \tag{5.120}$$

The probability associated with the wavefunction 5.120 is:

$$\psi^*(x)\psi(x) = \frac{1}{\sqrt{2\pi a}} e^{-\frac{x^2}{2a}} \tag{5.121}$$

which is compared with the Gaussian:

$$F(x) = \frac{1}{\sqrt{2\pi\sigma_x^2}} e^{-\frac{x^2}{2\sigma_x^2}} \tag{5.122}$$

giving $\sigma_x = \sqrt{a}$. It follows that:

$$\sigma_x\sigma_k = \frac{\sqrt{a}}{2\sqrt{a}} = \frac{1}{2}. \tag{5.123}$$

It can be shown that, using de Broglie relation $p = \hbar k$, we have $\sigma_p = \hbar\sigma_k$ and eq. 5.123 becomes the HUP for our wavepacket, i.e. $\sigma_x\sigma_k = \frac{\sqrt{a}}{2\sqrt{a}} = \frac{\hbar}{2}$.

It seems that a wavepacket might be able to represent the quantum behavior of a particle. We have also seen that measurable quantities are associated with operators, in particular with their expectation values. Let us study the time dependence of the expectation values of operators. In general, we can write for an operator $\hat{A}$ operating on a wavefunction ψ:

$$\frac{d}{dt}\langle\hat{A}\rangle = \frac{d}{dt}\int \psi^* \hat{A} \psi dx. \tag{5.124}$$

If we bring the derivative inside the integral sign, we have:

$$\frac{d}{dt}\langle\hat{A}\rangle = \int \frac{\partial\psi^*}{\partial t} \hat{A} \psi + \int \psi^* \frac{\partial\hat{A}}{\partial t}\psi + \int \psi\hat{A}\frac{\partial\psi}{\partial t} \tag{5.125}$$

where, for simplicity we are not writing explicitly the "dx" in the integral. Schrödinger equation 5.94 tells us that the Hamiltonian operator $\hat{H}$ is formally identified with the operator $i\hbar\frac{\partial}{\partial t}$. Eq. 5.125 becomes:

$$\begin{aligned}
\frac{d}{dt}\langle\hat{A}\rangle &= -\frac{i}{\hbar}\int \psi^*\hat{A}\hat{H}\psi + \frac{i}{\hbar}\int \psi^*\hat{H}\hat{A}\psi + \int \psi^*\frac{\partial\hat{A}}{\partial t}\psi \\
&= \frac{i}{\hbar}\int \psi^*[\hat{H},\hat{A}]\psi + \int \psi^*\frac{\partial\hat{A}}{\partial t}\psi.
\end{aligned} \tag{5.126}$$

The above equation becomes simpler if the operator $\hat{A}$ does not explicitly depend on the time. In this case, eq. 5.126 becomes:

$$\frac{d}{dt}\langle\hat{A}\rangle = \frac{i}{\hbar}\int\psi^*[\hat{H},\hat{A}]\psi.$$
(5.127)

Eq. 5.127 expresses a very important result: the expectation value of operators that commute with the Hamiltonian are constant of motion. This can be easily seen by putting $[\hat{H},\hat{A}] = 0$ in eq. 5.127.

Let's conclude this section by verifying that classical mechanics is contained in quantum physics if we associate expectation values of operators with classical observables. In the case of the operator $\hat{x}$ representing a free particle of mass m, for the commutator we have:

$$[\hat{H},x] = [\frac{\hat{p}^2}{2m},\hat{x}] = \frac{1}{2m}[\hat{p}\hat{p},\hat{x}] = \frac{1}{2m}(\hat{p}[\hat{p},\hat{x}] + [\hat{p},\hat{x}]\hat{p}) = -i\hbar\hat{p}$$
(5.128)

and eq. 5.127 can be written as:

$$\frac{d\langle\hat{x}\rangle}{dt} = \frac{i}{\hbar}\int\psi^*[\hat{H},\hat{x}]\psi = \frac{1}{m}\int\psi^*\langle\hat{p}\rangle\psi$$
(5.129)

which is equivalent to the classical mechanics relation $x = \frac{p}{m}$. Eq. 5.129 is referred to as *Ehrenfest's theorem*. For the momentum $\langle\hat{p}\rangle$ it is easy to verify that:

$$\frac{d\langle\hat{p}\rangle}{dt} = -\int\psi^*\langle\left(\frac{dV}{dx}\right)\rangle\psi$$
(5.130)

which is the equivalent of Newton's law $F = \frac{dp}{dt} = -\frac{dV}{dx}$.

5.6   TIME-INDEPENDENT SCHRÖDINGER EQUATION

Schrödinger equation can be simplified if the potential energy $V(r)$ does not depend on the time t. We can attempt a solution of the form:

$$\psi(r,t) = u(r)T(t)$$
(5.131)

where $u(r)$ depends only on the spatial coordinate r and $T(t)$ depends only on the time t. Inserting the trial solution 5.131 into Schrödinger equation 5.54 we have:

$$\frac{i\hbar}{T}\frac{dT}{dt} = \frac{1}{\psi}\left(-\frac{\hbar^2}{2m}\nabla^2 + V\right)\psi.$$
(5.132)

We notice that the left-hand side of eq. 5.132 depends only on t while the right-hand side depends only on r. This means that we can equate each of the sides of eq. 5.132 to the same constant E, the total energy, to give two independent equations. The first equation is:

$$i\hbar\frac{dT}{dt} = ET$$
(5.133)

which is easily integrated to give:

$$T(t) = Ce^{-\frac{iEt}{\hbar}}. \tag{5.134}$$

The second equation is:

$$\left(-\frac{\hbar^2}{2m}\nabla^2 R + VR\right)u(r) = Eu(r) \tag{5.135}$$

or, equivalently, in terms of the Hamiltonian operator acting on the wavefunction $u(r)$:

$$\hat{H}u(r) = Eu(r) \tag{5.136}$$

where it is important to notice that $\hat{H}$ is the Hamiltonian operator and E is a constant, in this case the energy. Eq. 5.136 is referred as *time independent Schrödinger equation* (TISE) and is used to describe stationary (with time) physical systems, i.e. systems for which the energy is conserved.

The class of equations of the form 5.136 are called *eigenvalue*[14] *equations* where the constant E is the *eigenvalue* while the function $u(r)$ is the *eigenfunction*. Therefore, solving the TISE is equivalent to find the eigenvalues and the eigenfunctions of the Hamiltonian operator $\hat{H}$.

The special class of wavefunctions solution of the TISE have a special property: expectation values and probability densities are independent of time. In fact, we have that the integrands containing $\psi^* \psi$ become:

$$\begin{aligned}
|\psi(r,t)|^2 &= \psi^*(r,t)\psi(r,t) \\
&= [u(x)T(t)]^* [u(x)T(t)] \\
&\propto u^*(r)e^{\frac{iEt}{\hbar}}u(r)\psi(r)e^{-\frac{iEt}{\hbar}} \\
&= u^*(r)u(r)
\end{aligned} \tag{5.137}$$

More in general, eq. 5.136 is valid for a set of values of the eigenvalues E_i with corresponding eigenvectors $u_i(r)$. If the eigenvalues form a discrete set ($i = 1,...,N$) then the quantum system is in a bound state[15] like, for example, the hydrogen atom with its discrete energy levels. If the system is unbound, then the eigenvalues are described by a continuous range of energies, like for example a free particle.

In the case of a bound state, eq. 5.136 becomes:

$$\hat{H}u_i(r) = E_i u(r) \tag{5.138}$$

Although discrete, eigenvalues and eigenfunctions in 5.138 can be infinite in number.

---

[14] The prefix *eigen* comes from German and means "own", "characteristic", "special" and other similar concepts.

[15] A bound state can be represented by a particle forced to remain localized in one or more regions of space. This can be achieved, for example, by the potential due to the presence of another particle. In this specific case the interaction energy is greater than the energy needed to separate the two particles.

We now ask whether eq. 5.138 is properly describing the energy spectrum of a bounded system. In order to do so, we need to make a few assumptions. If the operator $\hat{H}$ is to represent the energy, then its eigenvalues must *all* be real numbers because they represent the result of measurements of the energy. We also need to assume that the spectrum of eigenvalues is complete, i.e. the set of E_i contains all possible outcome of the measurements. Obviously, the set of E_i cannot contain additional values that cannot be obtained by measurements.

We see that to each energy eigenvalue E_i corresponds a specific wavefunction, called **eigenfunction** $u_i(r)$ that we now indicate with $\psi_i(r)$. We have seen that, according to the Copenhagen interpretation, the system *before* making a measurement of the energy can be in any of the states E_i with corresponding eigenfunction $\psi_i(r)$. However, as shown in fig. 5.9, *after* the process of measurement has been completed we know that the system is in the state corresponding to a specific eigenvalue of energy, let's say E_k. The act of measurement has had, as a consequence, that we now know the wavefunction of the system to be the corresponding eigenfunction. We say that the wavefunction of the system has collapsed to the eigenfunction $\psi_k(r)$.

We will see later that, given the **linearity** of the TISE and the **completeness** of the eigenfunctions, the wavefunction of the system, *before* making a measurement, is described by the sum of all the eigenfunctions:

$$\psi(x) = \sum_i e_i \psi_i(x) \tag{5.139}$$

where the e_i are normally complex coefficients. As we already mentioned, after the measurement the wavefunction has collapsed into the eigenfunction corresponding to the eigenvalue measured.

5.7 HERMITIAN OPERATORS

Let us now continue in the investigation of more interesting properties of operators. We now introduce the **Hermitian** operators. We already mentioned the linearity of Schrödinger equation but we did not specify the details.

An operator $\hat{A}$ is **linear** if it satisfy the following two conditions:

$$\begin{aligned} 1: \quad & \hat{A}(u(x) + v(x)) = \hat{A}u(x) + \hat{A}v(x) \\ 2: \quad & \hat{A}(c \cdot u(x)) = c\hat{A}u(x). \end{aligned} \tag{5.140}$$

So far we have assumed that operators always act on the object that is on their right as stated, for example, in the Schrödinger equation where the wavefunction $\psi(x, t)$ is on the right of the energy operator $-i\hbar\frac{\partial}{\partial t}$ or the Hamiltonian operator $\hat{H}$. We have seen that the expectation value of an operator is usually expressed as a "sandwich" structure $\psi^*\hat{A}\psi$ between the complex conjugate of the wavefunction ψ^* and the wavefunction ψ inside the integral like, for example, in eq. 5.89. The order in which we evaluate the integral is the following: we first evaluate the new function $(\hat{A}\psi)$ and then we multiply by the complex conjugate of ψ to obtain $\psi^*\hat{A}\psi$.

Now we ask if we can interpret the sandwich operation $\psi^*\hat{A}\psi$ in a different way by assuming that the operator $\hat{A}$ is now operating *on its left* on the complex conjugate function ψ^*. In order to indicate that the operator is now acting on its left, we write it as $\hat{A}^\dagger$ so that now the sandwich structure can be calculated as $(\psi^*\hat{A}^\dagger)\psi$. The operator $\hat{A}^\dagger$ is called the *adjoint* of the operator $\hat{A}$ and it can be different from the original operator $\hat{A}$. The adjoint of the operator $\hat{A}$ is defined by:

$$\int \psi^*\hat{A}\psi \, d\tau = \int \left(\psi^*\hat{A}^\dagger\right)^* \psi \, d\tau. \tag{5.141}$$

In quantum mechanics those operators that we want to represent physical quantities must have real eigenvalues. In this way we are guaranteed that every time we make a measurement on the system we obtain a real value. If $\hat{A}$ is one of such operators, this means that we want its expectation value to be a real number. This condition requires that $\langle\hat{A}\rangle = \langle\hat{A}\rangle^*$. Therefore:

$$\langle\hat{A}\rangle = \int \psi^*\hat{A}\psi \, d\tau = \left(\int \psi^*\hat{A}\psi \, d\tau\right)^* \tag{5.142}$$

We now assume that the operation of complex conjugation when applied to an operator acting on a function is $(\hat{A}\psi)^* = \psi^*\hat{A}^\dagger$, i.e. complex conjugation changes the operator to act on its left and on the complex conjugate of the original function. With this assumption, equation 5.142 becomes:

$$\langle\hat{A}\rangle = \int \psi^*\hat{A}\psi \, d\tau = \int \psi^*\hat{A}^\dagger\psi \, d\tau \tag{5.143}$$

and the condition of real eigenvalues is satisfied if the operator is such that:

$$\hat{A} = \hat{A}^\dagger. \tag{5.144}$$

An operator satisfying the condition 5.144 is called **Hermitian** or *self-adjoint*[16]. It is clear therefore that a Hermitian operator implies that it can operate indifferently on the function both on its right or on its left.

It can be easily seen that the Hermitian operator definition gives us a rule to calculate complex conjugates of the sandwich $\psi^*\hat{A}\psi$:

$$\left(\int \psi^*\hat{A}\psi \, d\tau\right)^* = \int \left(\psi^*\hat{A}\psi \, d\tau\right)^* = \int (\hat{A}\psi)^*\psi \, d\tau. \tag{5.145}$$

Let's verify that the position operators $\hat{x}$ is Hermitian. We need to prove that $\langle\hat{x}\rangle = \langle\hat{x}\rangle^*$. Using eq. 5.145, we have:

$$\langle\hat{x}\rangle^* = \int (\hat{x}\psi)^*\psi \, dx = \int (x\psi)^*\psi \, dx = \int \psi^* x\psi \, dx = \int \psi^*\hat{x}\psi \, dx = \langle\hat{x}\rangle \tag{5.146}$$

---

[16]Using "Hermitian" and "self-adjoint" indifferently is a convention among quantum physicists. Mathematicians might (and will) disagree.

where we used (twice) the fact that the position operator acting on a wavefunction is equal to the multiplication of the wavefunction by the coordinate x, i.e. $\hat{x}\psi = x\psi$. We also use the fact that the coordinates x are real and therefore $x^* = x$.

Let's verify that the momentum operator $\hat{p} = -i\hbar\frac{d}{dx}$ is also Hermitian. Again, we need to verify that $\langle\hat{p}\rangle = \langle\hat{p}\rangle^*$. We have:

$$\langle\hat{p}\rangle^* = \int (\hat{p}\psi)^*\psi\, dx = \int \left(-i\hbar\frac{d\psi}{dx}\right)^*\psi\, dx = i\hbar\int \frac{d\psi^*}{dx}\psi\, dx. \qquad (5.147)$$

We now integrate by part the last integrand of eq. 5.147. Let's refresh the integration by part rule: given two functions u and v, we have that:

$$\int_a^b \frac{du}{dx}v\,dx = uv\big|_a^b - \int_a^b u\frac{dv}{dx}dx. \qquad (5.148)$$

We now use the rule 5.148 to eq. 5.147:

$$\langle\hat{p}\rangle^* = i\hbar\int \frac{d\psi^*}{dx}\psi\, dx = -i\hbar\int \psi^*\frac{d\psi}{dx}\,dx = \langle\hat{p}\rangle. \qquad (5.149)$$

Notice that the proof above requires that the wavefunction goes to zero at the extremes of the interval of definition of the variable x so that the evaluated term $uv\big|_a^b = 0$.

It can be shown easily that the Hamiltonian operator is Hermitian and in general, all operators representing physical quantities are Hermitian like, for example, kinetic energy, potential energy, angular momentum, etc.

Let us now study briefly a few examples of the usage of the TISE for some simple quantum systems.

5.8 FREE PARTICLE

A free particle is apparently the conceptually simplest system to study because the TISE will reduce to its simplest form. However, as we will see in the following, the quantum mechanical treatment shows that the solutions are characterized by some interesting and important properties.

A free particle in one dimension x, i.e. a particle whose Hamiltonian has the potential $V(x) = 0$, is represented by the following Schrödinger (TISE) equation:

$$-\frac{\hbar^2}{2m}\frac{d^2\psi(x)}{dx^2} = E\psi(x) \qquad (5.150)$$

where we can use the total derivative since $\psi = \psi(x)$ does not depend on time or the other spatial coordinates y, z. If we call $k^2 = \frac{2mE}{\hbar^2}$ eq. 5.150 becomes:

$$\frac{d^2\psi(x)}{dx^2} + k^2\psi(x) = 0. \qquad (5.151)$$

A solution in exponential form of eq. 5.151 can be written as:

$$\psi(x) = A_1 e^{ikx} + A_2 e^{-ikx} \tag{5.152}$$

and, using eq. 5.134, the time-dependent wavefunction solution is:

$$\psi(x,t) = A_1 e^{ik(x - \frac{\hbar k}{2m}t)} + A_2 e^{-ik(x + \frac{\hbar k}{2m}t)} \tag{5.153}$$

where we used $e^{-\frac{iEt}{\hbar}} = e^{-\frac{\hbar k^2}{2m}}$.

It is interesting to note the argument of the exponentials in 5.153:

$$x \mp \frac{\hbar k}{2m} t = x \mp v_p t \tag{5.154}$$

indicating that the wavefunction 5.153 is a stationary state obtained by the superposition of two waves of fixed shape, moving, respectively, to the right with velocity $+v_p = +\frac{1}{2}\frac{\hbar k}{m}$ and to the left with velocity $-v_p = -\frac{1}{2}\frac{\hbar k}{m}$. Let's compare the velocity v_p with the classical velocity obtained by $v_c = \frac{p}{m} = \frac{\hbar k}{m} = 2v_p$. The speed of the wavefunction shape, i.e. the phase velocity $v_p = \frac{\omega}{k}$ is half the speed of the classical particle! The group velocity, on the other hand, $v_g = \frac{d\omega}{dk} = v_c$ is equal to the speed of the particle, as expected.

We can write the wavefunction 5.153 in a compact form when the two coefficients are equal, i.e. $A_1 = A_2$:

$$\psi(x,t) = A e^{i(kx - \frac{\hbar k^2}{2m}t)}. \tag{5.155}$$

Let's check the normalization of the wavefunction 5.155:

$$\int_{-\infty}^{+\infty} \psi^* \psi \, dx = |\hat{A}|^2 \int_{-\infty}^{+\infty} dx \to \infty. \tag{5.156}$$

We have the remarkable result that the wavefunction of a free particle cannot be normalized, and therefore, does not represent physical states. Since we know that free particles happily exist we can still build a general solution by superimposing many stationary states which is what we have done when we discussed the wavepackets in section 5.5.

Since there is no condition on the values of the wavevector k, then we can safely assume that it is a continuous variable and therefore we can build a wavepacket by integrating over a continuous spectrum of stationary states:

$$\psi(x,t) = A \int_{-\infty}^{+\infty} \phi(k) e^{i(kx - \frac{\hbar k^2}{2m}t)} \, dk \tag{5.157}$$

where the function $\phi(k)$ represents the relative weight to be associated to each exponential term in analogy[17] to eq. 5.113 of section 5.5.

---

[17] In eq. 5.157, we have included the time dependent term not included in eq. 5.113 of section 5.5.

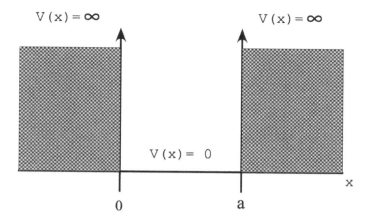

Figure 5.12   Potential $V(x)$ of a particle bound on a segment $(0, a)$.

5.9   PARTICLE IN A BOX

The particle in a box, or more accurately on a segment since we will study only the x-dependence, is an idealization of a particle confined between bounds that are impenetrable or infinitely rigid. We assume that a particle of mass m is bound to live on a segment because of a potential which is $V(x)$ zero when $0 \leq x \leq a$ and is infinite outside the segment $(0, a)$. The potential therefore is such that:

$$V(x) = \begin{cases} 0 & 0 \leq x \leq a \\ \infty & x < 0 \text{ and } x > a \end{cases} \tag{5.158}$$

The particle can move only in one dimension x and the TISE equation describing this system is:

$$-\frac{\hbar^2}{2m}\frac{d^2\psi}{dx^2} + V(x) = E(x) \tag{5.159}$$

The potential $V(x)$, represented in fig. 5.12 imposes some conditions on the wavefunction $\psi(x)$. First, the wavefunction must be zero in the region where the potential is infinite, i.e. $\psi(x) = 0$ for $0 \leq x \leq a$ reflecting that fact that it is impossible to find the particle in the forbidden region. In fact, a particle cannot be in this forbidden region because it would be required to have infinite energy. The particle is said to be in a *potential well*. The second condition requires that $\psi(x) = 0$ for $x = 0$ and $x = a$. This condition is required if we want to have the wavefunction to be continuous over the entire x-axis $-\infty < x < \infty$.

With these boundary conditions the TISE 5.159 simplifies to:

$$-\frac{\hbar^2}{2m}\frac{d^2\psi}{dx^2} = E(x) \tag{5.160}$$

which is exactly identical to eq. 5.151 if we use the same substitution:

$$k^2 = \frac{2mE}{\hbar^2}. \tag{5.161}$$

The most general solution is:

$$\psi(x) = A\sin kx + B\cos kx. \tag{5.162}$$

The first boundary condition $\psi(0) = 0$ gives:

$$\psi(0) = A\sin 0 + B\cos 0 = 0 \rightarrow B = 0 \tag{5.163}$$

where we assume that, since $B = 0$, then $A \neq 0$. The second boundary condition $\psi(a) = 0$ gives:

$$\psi(a) = A\sin ka = 0 \rightarrow ka = n\pi \tag{5.164}$$

where n is an integer. Therefore, we have a set of n wavefunctions all solution of eq. 5.160:

$$\psi_n(x) = A\sin\frac{n\pi}{a}x. \tag{5.165}$$

If we now insert $k = \frac{n\pi}{a}$ in eq. 5.161 we have that the energy is:

$$E_n = \frac{\pi^2\hbar^2}{2ma^2}n^2, \quad n = 1,2,3,... \tag{5.166}$$

where we indicated the energy E_n with a subscript n to point out that the energy is quantized. The particle can only have a discrete set of allowed energies according to eq. 5.166 in contrast with the case of a classical particle with the same potential.

While a classical particle can happily be standing still or moving with any possible energy inside the segment $(0,a)$ with a continuous spectrum, a quantum particle is subject to a restriction on the possible value of energies E_n given by eq. 5.166.

Perhaps the most remarkable feature consists in the fact that **the particle is not allowed to have zero energy**. In fact, the particle must possess a minimum energy $\frac{\pi^2\hbar^2}{2ma^2}$, corresponding to $n = 1$[18]. In this case the particle is said to be in the *ground state* or, alternatively, we say that the particle has *zero point energy*.

A classical particle standing still will have $E = 0$ which would correspond to knowing *exactly* where the particle is located on the segment, therefore having $\Delta x = 0$. But we know that, according to the HUP, there would be maximum uncertainty in the momentum Δp. The zero point energy is therefore consistent with the HUP $\Delta x\Delta p \sim h$. We can see this by considering that there is a minimum amount of kinetic energy for the quantum particle corresponding to:

$$E_1 = \frac{\pi^2\hbar^2}{2ma^2} = \frac{p^2}{2m} \tag{5.167}$$

---

[18] The case $n = 0$ would correspond to a wavefunction $\psi(x) = $ constant which cannot be normalized to represent a physical system.

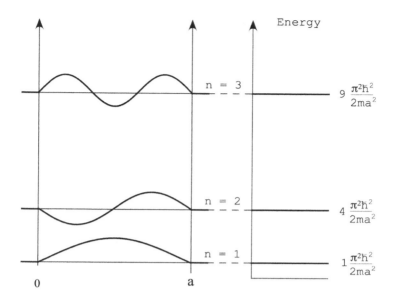

Figure 5.13 Wavefunctions and energy levels of a particle in a box.

which gives a $p_{min} = \pm \frac{\pi \hbar}{a}$ which can be interpreted as an uncertainty in the momentum of the order of $\Delta p \sim \frac{2\pi \hbar}{a}$. If we now assume that the uncertainty in the position of the particle on the segment is of the order of the size of the segment, $\Delta x \sim a$, we have that:

$$\Delta x \Delta p = a \cdot \frac{2\pi \hbar}{a} \; 2\pi \hbar \sim h. \tag{5.168}$$

showing the consistency with the HUP.

Let us go back to the eigenfunctions 5.165 and let's normalize them, i.e. let's find the value of the coefficient A such that:

$$\int_{-\infty}^{+\infty} |\psi_n(x)|^2 dx = \int_0^a |\psi_n(x)|^2 dx = 1. \tag{5.169}$$

We have that:

$$\int_0^a |\psi_n(x)|^2 dx = A^2 \int_0^a \sin^2 \frac{n\pi x}{a} dx = 1. \tag{5.170}$$

Let's now make the substitution $\alpha = \frac{n\pi x}{a}$ with $dx = \frac{a}{n\pi} d\alpha$ and $d\alpha$, running between 0 and $n\pi$, in eq. 5.170:

$$A^2 \int_0^a \sin^2 \frac{n\pi x}{a} dx = \frac{aA^2}{n\pi} \int_0^{n\pi} \sin^2 \alpha \, d\alpha = \frac{aA^2}{n\pi} \cdot \frac{n\pi}{2} = \frac{aA^2}{2} = 1 \tag{5.171}$$

from which we obtain the normalized form of the eigenfunctions:

$$\psi_n(x) = \sqrt{\frac{2}{a}} \sin \frac{n\pi}{a} x \tag{5.172}$$

where $n = 1, 2, 3, \ldots$.

Another remarkable feature of the eigenfunctions of the particle in a box consists in their **orthonormality** expressed through the study of the integral of the product of two different eigenfunctions.

Let's show that the following integral is always zero for $m \neq n$:

$$\int_{-\infty}^{+\infty} \psi_m(x)\psi_n(x)dx = 0, \text{ for } m \neq n \tag{5.173}$$

Using the trigonometric identity:

$$\sin \alpha \sin \beta = \frac{1}{2}[\cos(\alpha - \beta) - \cos(\alpha + \beta)] \tag{5.174}$$

eq. 5.173 becomes:

$$
\begin{aligned}
\int_{-\infty}^{+\infty} \psi_m(x)\psi_n(x)dx &= \frac{2}{a}\int_0^a \sin\left(\frac{m\pi}{a}x\right)\sin\left(\frac{n\pi}{a}x\right)dx \\
&= \frac{1}{a}\int_0^a \left[\cos[(m-n)\frac{\pi x}{a}] - \cos[(m+n)\frac{\pi x}{a}]\right]dx \\
&= -\frac{1}{a}\left[[\sin(m-n)\frac{\pi x}{a}]\right]_0^a + \frac{1}{a}\left[[\sin(m+n)\frac{\pi x}{a}]\right]_0^a \\
&= 0
\end{aligned}
\tag{5.175}
$$

and last equality is always valid because both $(m-n)$ and $(m+n)$, for m and n integers, are always integer causing the sin to be zero when evaluates between 0 and a.

We can include the normalization condition of the eigenfunctions to rewrite eq. 5.173 as:

$$\int_{-\infty}^{+\infty} \psi_m(x)\psi_n(x)dx = \int_0^a \psi_m(x)\psi_n(x)dx = \delta_{mn} \tag{5.176}$$

where the symbol δ_{mn}, or Kronecker delta, is a function of the two integer indexes mn such that:

$$\delta_{mn} = \begin{cases} 0 & m \neq n \\ 1 & m = n. \end{cases} \tag{5.177}$$

5.10 THE HARMONIC OSCILLATOR

The simple (linear) harmonic oscillator is of paramount importance in classical as well as quantum mechanics. We have seen, for example, that the very birth of quantum mechanics required the use of harmonic oscillators in a cavity to describe the e.m. radiation spectrum which led to Planck's hypothesis.

Let's assume, for simplicity, that the system is composed of a mass m attached to a wall by a spring of elastic constant η. If the mass is constrained to move only along the x-axis, then Hooke's law states that the force on the mass m is $F = -\eta x$. Notice that usually the elastic constant is indicated with k. However, in order to avoid confusion with the wave vector k, we prefer to indicate the elastic constant in Hooke's law with the symbol η.

Newton's law can be written as:

$$-\eta x = m\frac{d^2 x}{dt^2} \tag{5.178}$$

which has the solution:

$$x(t) = A\sin(\omega t) + B\cos(\omega t) \tag{5.179}$$

where $\omega = \sqrt{\frac{\eta}{m}}$ is the classical frequency of the oscillator. The potential energy is:

$$V(x) = \frac{1}{2}\eta x^2 = \frac{1}{2}m\omega^2 x^2. \tag{5.180}$$

which is the well known classical result.

5.10.1   ALGEBRAIC SOLUTION

Let's now study the quantum version of the harmonic oscillator. This is achieved by by writing down the Schrödinger equation (TISE) with the classical potential 5.180:

$$-\frac{\hbar^2}{2m}\frac{d^2\psi}{dx^2} + \frac{1}{2}m\omega^2 x^2\psi = E\psi. \tag{5.181}$$

Let's write the Hamiltonian of the harmonic oscillator highlighting the quantum operators $\hat{p}$ and $\hat{x}$:

$$\hat{H} = \frac{1}{2m}\left(\hat{p}^2 + m^2\omega^2\hat{x}^2\right). \tag{5.182}$$

We now show that we can learn a lot about the quantum harmonic oscillator without explicitly solving Schrödinger equation, by introducing two new operators[19] a_+, a_-.

Inspired by the fact that we can factorize the sum of the squares of two real numbers a and b as $a^2 + b^2 = (a + ib)(a - ib)$, we might try factorizing the Hamiltonian 5.182 in a similar way by introducing two new operators:

$$a_+ = \frac{1}{\sqrt{2\hbar m\omega}}(-i\hat{p} + m\omega\hat{x})$$

$$\tag{5.183}$$

$$a_- = \frac{1}{\sqrt{2\hbar m\omega}}(i\hat{p} + m\omega\hat{x}).$$

The normalization constant $\frac{1}{\sqrt{2\hbar m\omega}}$ is chosen for convenience as we will see later. The two operators 5.183 are called *ladder operators* or *raising* and *lowering* operators, respectively, for a_+ and a_-[20]. Notice that these two operators are Hermitian conjugates but not Hermitian, i.e. $a_- = (a_+)^\dagger \neq (a_-)^\dagger$.

---

[19] This method was first used by P.A.M. Dirac.

[20] Often the terms *creation* and *annihilation* operators are used too.

Let's calculate the product of the two operators $a_- a_+$:

$$
\begin{aligned}
a_- a_+ &= \frac{1}{2\hbar m\omega}(i\hat{p} + m\omega\hat{x})(-i\hat{p} + m\omega\hat{x}) \\
&= \frac{1}{2\hbar m\omega}(\hat{p}^2 + m^2\omega^2\hat{x}^2 - im\omega[\hat{x},\hat{p}]) \\
&= \frac{1}{2\hbar m\omega}(\hat{p}^2 + m^2\omega^2\hat{x}^2 + \hbar m\omega)
\end{aligned}
\tag{5.184}
$$

which can be written in terms of the Hamiltonian of the harmonic oscillator as:

$$
a_- a_+ = \frac{\hat{H}}{\hbar\omega} + \frac{1}{2}.
\tag{5.185}
$$

Similar calculation will give for $a_+ a_-$:

$$
\begin{aligned}
a_- a_+ &= \frac{1}{2\hbar m\omega}(-i\hat{p} + m\omega\hat{x})(i\hat{p} + m\omega\hat{x}) \\
&= \frac{1}{2\hbar m\omega}(\hat{p}^2 + m^2\omega^2\hat{x}^2 - im\omega[\hat{p},\hat{x}]) \\
&= \frac{1}{2\hbar m\omega}(\hat{p}^2 + m^2\omega^2\hat{x}^2 - \hbar m\omega)
\end{aligned}
\tag{5.186}
$$

which can also be written in terms of the Hamiltonian as:

$$
a_+ a_- = \frac{\hat{H}}{\hbar\omega} - \frac{1}{2}.
\tag{5.187}
$$

Equations 5.185 and 5.187 allow us to immediately calculate the commutator:

$$
[a_+, a_-] = 1
\tag{5.188}
$$

and an expression for the Hamiltonian:

$$
\hat{H} = \frac{1}{2}\hbar\omega\,(a_- a_+ + a_+ a_-).
\tag{5.189}
$$

Depending on which product of ladder operators we use, Schrödinger TISE can be written as:

$$
\begin{aligned}
\hbar\omega\left(a_+ a_- + \frac{1}{2}\right)\psi = E\psi \\
\hbar\omega\left(a_- a_+ - \frac{1}{2}\right)\psi = E\psi.
\end{aligned}
\tag{5.190}
$$

Let's study the action of the raising and lowering operators on the wavefunction ψ satisfying the TISE for harmonic oscillators. For the raising operator $a_+\psi$ we have:

$$
\begin{aligned}
\hat{H}(a_+\psi) &= \hbar\omega\left(a_+ a_- + \frac{1}{2}\right)(a_+\psi) \\
&= \hbar\omega\left(a_+ a_- a_+ + \frac{1}{2}a_+\right)\psi \\
&= \hbar\omega a_+\left(a_- a_+ + \frac{1}{2}\right)\psi.
\end{aligned}
\tag{5.191}
$$

Substituting back $a_- a_+$ from eq. 5.185, we have:

$$\hat{H}(a_+ \psi) = \hbar \omega a_+ \left(a_- a_+ + \frac{1}{2} \right) \psi$$

$$= \hbar \omega a_+ \left(\frac{\hat{H}}{\hbar \omega} + 1 \right) \psi \qquad (5.192)$$

$$= a_+ \left(\hat{H} + \hbar \omega \right) \psi$$

$$= (E + \hbar \omega) a_+ \psi.$$

Eq. 5.192 tells us that, if ψ is a solution to the Schrödinger equation (TISE) corresponding to the eigenvalue E, then the new wavefunction $a_+ \psi$ is also solution to the Schrödinger equation (TISE) corresponding to the eigenvalue $(E + \hbar \omega)$ thus justifying the name "raising" operator because it raises the energy by a unit of $\hbar \omega$. Similarly it can be shown that:

$$\hat{H}(a_- \psi) = (E - \hbar \omega) a_- \psi \qquad (5.193)$$

justifying the name of "lowering" operator because it lowers the energy by a unit of $\hbar \omega$.

Let's study the following expectation value:

$$\langle a_+ a_- \rangle = \int \psi^* (a_+ a_-) \psi \, d\tau. \qquad (5.194)$$

And let's call $\psi' = a_- \psi$. Being a_+ the Hermitian conjugate of a_-, it follows that $\psi'^* = \psi^* a_+$. therefore we have that:

$$\langle a_+ a_- \rangle = \int \psi^* (a_+ a_-) \psi \, d\tau = \int \psi'^* \psi' \, d\tau \geq 0. \qquad (5.195)$$

Using eq. 5.187 we have:

$$\langle a_+ a_- \rangle = \int \psi^* (a_+ a_-) \psi \, d\tau = \int \psi^* \left(\frac{\hat{H}}{\hbar \omega} - \frac{1}{2} \right) \psi \, d\tau = \left(\frac{E}{\hbar \omega} - \frac{1}{2} \right) \int \psi^* \psi \, d\tau$$

$$(5.196)$$

and in virtue of eq. 5.195 we must have that:

$$\left(\frac{E}{\hbar \omega} - \frac{1}{2} \right) \geq 0 \rightarrow E \geq \frac{1}{2} \hbar \omega \qquad (5.197)$$

Eq. 5.197 tells us that the minimum energy allowed to the quantum oscillator is not zero but $E \geq \frac{1}{2} \hbar \omega$. Therefore, in analogy of the particle constrained on a segment, we see that also the quantum oscillator has a minimum nonzero energy, i.e. a zero point energy equal to $\frac{1}{2} \hbar \omega$. States with lower energy are not allowed. This means that we can only have energies bigger than the minimum energy $\frac{1}{2} \hbar \omega$.

The question is: is the spectrum of the harmonic oscillator continuous or discrete? We already know the answer which was provided by Planck. The whole quantum mechanics has its foundation on the hypothesis of quanta $E = \hbar\omega$ and therefore there can only be discrete energy levels separated by an energy $\Delta E = \hbar\omega$.

In order to build the energy spectrum of the quantum harmonic oscillator we just start from the lowest energy level and assign to the eigenfunction ψ_0 the corresponding eigenvalue $E_0 = \frac{1}{2}\hbar\omega$ so that Schrödinger TISE is $\hat{H}\psi_0 = E_0\psi_0$.

Let's now apply the operator a_+ to the eigenfunction ψ_0. Using eq. 5.192 we have:

$$\hat{H}(a_+\psi_0) = (E_0 + \hbar\omega)(a_+\psi_0) = \left(\frac{1}{2}\hbar\omega + \hbar\omega\right)(a_+\psi_0) \qquad (5.198)$$

Eq. 5.198 tells us that the new wavefunction $\psi_1 = a_+\psi_0$ is an eigenfunction with eigenvalue $\left(\frac{1}{2}\hbar\omega + \hbar\omega\right)$. If we keep doing this operation, we find that we can write the Schrödinger TISE for the nth eigenfunction as:

$$\hat{H}\psi_n = \left(\frac{1}{2}\hbar\omega + n\hbar\omega\right)\psi_n. \qquad (5.199)$$

which gives the expression for the spectrum of the quantum harmonic oscillator:

$$E = \left(\frac{1}{2}\hbar\omega + n\hbar\omega\right) = \left(n + \frac{1}{2}\right)\hbar\omega. \qquad (5.200)$$

Notice that the use of ladders operators has allowed us to find the discrete spectrum of the energy of the quantum harmonic oscillator without actually solving Schrödinger equation 5.181.

Having seen that the ladders operators effectively describe the increase/decrease of a quantum $\hbar\omega$ in the energy of the oscillator, it is natural to assume the following:

$$a_-\psi_n = \beta_n\psi_{n-1}$$
$$\psi_n^* a_+ = \psi_{n-1}\beta_n* \qquad (5.201)$$

expressing the "lowering" action of a_- with its Hermitian conjugate a_+ and where β_n and β_n^* are complex conjugate numbers. With these assumptions, we can now write the expectation value of the operator a_+a_-:

$$\langle a_+a_-\rangle = |\beta_n|^2 \int \psi_{n-1}^* \psi_{n-1}\, d\tau = |\beta_n|^2. \qquad (5.202)$$

Last equality in eq. 5.202 holds because the eigenfunctions are orthonormal. Using eq. 5.187 we can also write[21]:

$$\langle a_+a_-\rangle = \int \psi_n^* \left(\frac{\hat{H}}{\hbar\omega} - \frac{1}{2}\right)\psi_n\, d\tau = n + \frac{1}{2} - \frac{1}{2} = n \qquad (5.203)$$

---

[21]The fact that the expectation value of the operator a_+a_- is the number n is the reason why this operator is called *number* operator.

from which it follows that $|\beta_n|^2 = n$ or $\beta_n = \sqrt{n}$ and we can write:

$$a_-\psi_n = \sqrt{n}\psi_{n-1} \qquad (5.204)$$

and eq. 5.204 implies that $a_-\psi_0 = 0$, i.e. the operator a_- *annihilate* the quantum state ψ_0.

Following the same procedure, we assume that:

$$a_+\psi_n = \gamma_n\psi_{n+1}$$
$$\psi_n^*a_- = \psi_{n+1}\gamma_n* \qquad (5.205)$$

where γ_n and γ_n^* are again complex conjugate numbers. The expectation value of the operator a_-a_+ is:

$$\langle a_-a_+ \rangle = |\gamma_n|^2 \int \psi_{n+1}^*\psi_{n+1}\, d\tau = |\gamma_n|^2. \qquad (5.206)$$

and using eq. 5.185, after a little algebra, we obtain that $\gamma_n = \sqrt{n+1}$ and therefore:

$$a_+\psi_n = \sqrt{n+1}\psi_{n+1} \qquad (5.207)$$

which justify the name of *creation* operator.

Having found the eigenvalues, let's find out the form of the eigenfunctions. Again, the ladders operators can help us find all the eigenfunctions once we know at least one. Let's start from $a_-\psi_0 = 0$ which can be written as:

$$\frac{1}{\sqrt{2m\hbar\omega}}(ip + m\omega x)\psi_0 = \frac{1}{\sqrt{2m\hbar\omega}}\left(\hbar\frac{d}{dx} + m\omega x\right)\psi_0 = 0. \qquad (5.208)$$

Eq. 5.208 can be simplified to:

$$\frac{d}{dx}\psi_0 = -\frac{1}{\hbar}m\omega x\psi_0. \qquad (5.209)$$

which can be written as:

$$\frac{d\psi_0}{\psi_0} = -\frac{m\omega}{\hbar}x\, dx. \qquad (5.210)$$

which integrated gives:

$$\psi_0 = Ae^{-\frac{m\omega}{2\hbar}x^2}. \qquad (5.211)$$

Let's find the normalization constant A. We have that:

$$\begin{aligned}
|\psi_0|^2 &= |A|^2 \int_{-\infty}^{+\infty} \left(e^{-\frac{m\omega}{2\hbar}x^2}\right)^2 dx \\
&= |A|^2 \int_{-\infty}^{+\infty} e^{-\frac{m\omega}{\hbar}x}\, dx \\
&= |A|^2 \left(\frac{\pi\hbar}{m\omega}\right)^{\frac{1}{2}} \\
&= 1
\end{aligned} \qquad (5.212)$$

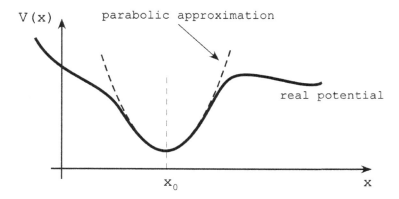

Figure 5.14 Quadratic (parabolic) approximation around a local minimum.

from which we finally obtain:

$$\psi_0 = \left(\frac{m\omega}{\pi\hbar} \right)^{\frac{1}{4}} e^{-\frac{m\omega x^2}{2\hbar}}. \tag{5.213}$$

Repeated application of the operator a_+ gives the other eigenfunctions:

$$\psi_n = \frac{1}{\sqrt{n!}} (a_+)^n \psi_0. \tag{5.214}$$

5.10.2 ANALYTICAL SOLUTION

A proper solution to the Harmonic oscillator problem can be found by solving Schrödinger equation analytically. Before going into the details of the calculation, let's discuss the potential of the harmonic oscillator. Hooke's force $F \propto -\eta x$, and relative Hooke's potential $V = \frac{1}{2}\eta x^2$, are usually idealization of real life devices. Hooke's potential, for example, goes to infinity for $x \to \pm\infty$ which is clearly not physical because it is impossible to build a spring that compresses or extends indefinitely with a linear force. However, if we limit ourselves to small oscillations where the linearity of Hooke's force is a good approximation then our physics is robust. It is important to notice that, no matter how complex the potential is, if there is a region where the potential has a *local* minimum, i.e. where $\frac{d^2 V}{dx^2} > 0$ then, as shown in fig. 5.14, around the minimum we can always expand the potential as a Taylor series:

$$V(x) = V_0 + V_1(x - x_0) + \frac{1}{2}V_2(x - x_0)^2 + \dots = \sum_{i=1}^{N} \frac{1}{i!}V_i(x - x_0)^i \tag{5.215}$$

where the V_i are the various derivatives calculated at $x = x_0$ and x_0 is the x-coordinate of the minimum. In our quadratic approximation, we only keep the first three terms: $V(x) = V_0 + V_1(x - x_0) + \frac{1}{2}V_2(x - x_0)^2$. The condition that at $x = x_0$ the potential has a minimum requires that the first derivative at x_0 is zero and therefore $V_1 = 0$. In addition, we can define the zero potential at $x = x_0$ so that $V_0 = 0$ thus justifying a potential of the form $V(x) = \frac{1}{2}\eta x^2$ where $\eta = V_2$.

Schrödinger equation with the harmonic oscillator potential is:

$$-\frac{\hbar^2}{2m}\frac{d^2\psi}{dx^2} + \frac{1}{2}\eta x^2\psi = E\psi. \qquad (5.216)$$

We already know that $\omega = \sqrt{\frac{\eta}{m}}$ and eq. 5.216 can be written as:

$$\frac{d^2\psi}{dx^2} + \left(\frac{2mE}{\hbar^2} - \frac{m\omega^2}{\hbar^2}x^2.\right)\psi = 0. \qquad (5.217)$$

For simplicity let's write $a = \frac{m\omega}{\hbar}$ and $b = \frac{2mE}{\hbar^2}$. Eq. 5.217 becomes:

$$\frac{d^2\psi}{dx^2} + \left(b - a^2 x^2\right)\psi = 0. \qquad (5.218)$$

It is a good practice to work with dimensionless equations. Eq. 5.217 contains the coefficients a and b which have dimensions of the inverse of the square of a length. A good choice for a change of variable would be $y = \sqrt{a}\,x$ so that the new variable y is now dimensionless.

Let's calculate the derivatives in the new variable. Applying the chain rule we have, for the first derivative:

$$\frac{d\psi}{dx} = \frac{d\psi}{dy}\frac{dy}{dx} = \frac{d\psi}{dy}\frac{d}{dx}(\sqrt{a}\,x)) = \sqrt{a}\,\frac{d\psi}{dy}. \qquad (5.219)$$

We can look at eq. 5.219 as a definition of the operator $\frac{d}{dx} = \sqrt{a}\,\frac{d}{dy}$. It follows that the second derivative is:

$$\frac{d^2}{dx^2} = \left(\sqrt{a}\,\frac{d}{dy}\right)\left(\sqrt{a}\,\frac{d}{dy}\right) = a\frac{d^2}{dy^2} \qquad (5.220)$$

and eq. 5.218 becomes:

$$\frac{d^2\psi}{dy^2} + \left(\frac{b}{a} - y^2\right)\psi = 0. \qquad (5.221)$$

We can verify that eq. 5.221 is dimensionless by inserting $\frac{b}{a} = \frac{2E}{\hbar\omega}$, which is obviously dimensionless, into eq. 5.221:

$$\frac{d^2\psi}{dy^2} + \left(\frac{2E}{\hbar\omega} - y^2\right)\psi = 0. \qquad (5.222)$$

Let's study what wavefunction satisfy eq. 5.221 (or equivalently 5.222) in the limit of $y \to \pm\infty$.

When $y^2 \gg E$ eq. 5.222 simplify to:

$$\frac{d^2\psi}{dy^2} - y^2\psi = 0. \tag{5.223}$$

We now impose that the wavefunction $\psi(y)$ is well behaved when $|y| \to \infty$, i.e. goes to zero together with its first derivative $\psi(y) \to 0$ and $\frac{d\psi}{dy} \to 0$ when $|y| \to \infty$. This will make sure that the wavefunction represents the probability amplitude of a physical state.

A good choice for the asymptotic solution is:

$$\psi_a(y) = Ae^{-\frac{y^2}{2}} \tag{5.224}$$

where the subscript a reminds us that this is the asymptotic solution. The next step would be to attempt writing the general solution as a product of the asymptotic solution $\psi_a(y)$ times a function $H(y)$ which will solve eq.5.223:

$$\psi(y) = \psi_a(y)H(y) = Ae^{-\frac{y^2}{2}}H(y). \tag{5.225}$$

In order to have a valid solution 5.225 we must be sure that $\psi_a(y)$ dominates for large $|y|$ while $H(y)$ dominates for small $|y|$. Inserting the trial solution 5.225 into 5.221, and after a lot of differentiations, we obtain a differential equation for the unknown function $H(y)$:

$$\frac{d^2H}{dy^2} - 2y\frac{dH}{dy} + \left(\frac{b}{a} - 1\right)H = 0. \tag{5.226}$$

Eq. 5.226 is well known in mathematical physics and is equivalent to the so called *Hermite* differential equation which has been first studied by the mathematician Charles Hermite[22].

The standard technique to solve differential equations like 5.226 consists of assuming that the solution can be written as a power series, i.e.:

$$H(y) = \sum_{n=1}^{\infty} c_n y^n = c_0 + c_1 y + c_2 Y^2 + c_3 y^3 + \dots \tag{5.227}$$

We now plug the expression 5.227 into eq. 5.226 and calculate each of its terms separately. When calculating the second derivative in the sum 5.227, only the terms $n \geq 2$ will be non zero. Therefore:

$$\frac{d^2H}{dy^2} = \sum_{n=0}^{\infty} (n+1)(n+2)c_{n+2}y^n. \tag{5.228}$$

---

[22]Charles Hermite (1822-1901) was a French mathematician who gave important contributions in many fields of mathematics used in physics like, for example, Hermitian operators and Hermite polynomials.

Similarly, for the first derivative term in eq. 5.226:

$$-2y\frac{dH}{dy} = -2\sum_{n=0}^{\infty} nc_n y^n. \tag{5.229}$$

The last term is simply:

$$\left(\frac{b}{a} - 1\right) H = \left(\frac{b}{a} - 1\right) \sum_{n=0}^{\infty} c_n y^n. \tag{5.230}$$

Putting eqs. 5.228, 5.229 and 5.230 together we have:

$$\sum_{n=0}^{\infty} \left[(n+2)(n+1)c_{n+2} - 2nc_n + \left(\frac{b}{a} - 1\right) c_n\right] y^n = 0 \tag{5.231}$$

which must hold for all values of y^n. Therefore we have that each individual coefficients of y^n in eq. 5.231 must be zero:

$$\left[(n+2)(n+1)c_{n+2} - 2nc_n + \left(\frac{b}{a} - 1\right) c_n\right] = 0. \tag{5.232}$$

Rearranging the terms, eq. 5.232 gives us a recursion relation allowing us to calculate all the coefficients c_n, for $n \geq 2$ of the power series 5.227. The coefficients c_0 and c_1 need to be determined from the initial conditions. The recursion relation is:

$$c_{n+2} = \frac{\left(2n+1-\frac{b}{a}\right)}{(n+2)(n+1)} c_n. \tag{5.233}$$

Given the structure of eq. 5.233 we see that starting with c_0 we obtain all the even terms while starting with c_1 we obtain all the odd terms. This means that we can write the function $H(y)$ as the sum of the even terms and the odd terms:

$$H(y) = H_{even}(y) + H_{odd}(y) \tag{5.234}$$

where:

$$\begin{aligned} H_{even}(y) &\equiv c_0 + c_2 y^2 + c_4 y^4 + ... \\ H_{odd}(y) &\equiv c_1 + c_3 y^3 + c_5 y^5 + ... \end{aligned} \tag{5.235}$$

Let's go back to the proposed solution 5.225 for the Quantum Harmonic Oscillator and check if it can be normalized[23]. Unfortunately it turns out that such a solution, with its infinite number of terms, cannot be normalized. The only chance we have to *force* a normalization would be to find a condition for which the series can be truncated or, in other words, we should find a maximum value of n, call it N, such

---

[23]For more details about the normalization of the solutions of the Quantum Harmonic Oscillator see, for example, the excellent book by Griffiths [19].

that the recursion formula gives $c_{n+2} = 0$. This condition will truncate either the even or the odd part of the expression 5.235.

Looking at the recursion relation 5.233, we see that the series can be truncated if we impose:

$$\left(\frac{b}{a}\right) = 2n + 1 \tag{5.236}$$

and remembering that $\left(\frac{b}{a}\right) = \frac{2E}{\hbar\omega}$, the normalization condition 5.236 gives us the quantization of the energy levels of the quantum harmonic oscillator we obtained algebraically previously:

$$E = \left(n + \frac{1}{2}\right)\hbar\omega. \tag{5.237}$$

For all this to be valid, we need to make the additional assumption that, if a maximum value $n = N$ that truncates the even series of the expression 5.235 is found, then we must assume that the odd series is zero from the beginning and vice versa.

In quantum mechanics, therefore, as a general rule, we find that if the system is bounded then its energy levels will be discrete once the wavefunction is properly normalized. In the Quantum Harmonic Oscillator example, the discretization relations 5.236 and 5.237 are clearly generated only if we impose the normalization condition on the wavefunction.

We finally have all the ingredients to calculate the eigenfunctions of the Quantum Harmonic Oscillator. To do so, we rewrite the recursion formula 5.233 inserting the relation 5.236 to obtain:

$$c_{n+2} = \frac{-2(N-n)}{(n+2)(n+1)}c_n. \tag{5.238}$$

The first eigenfunction will correspond to $N = 0$ with $n = 0$ since the integer n cannot be bigger than N. We know that we have to impose $c_1 = 0$ to make sure that the odd series in eq. 5.235 is zero. This choice generates just one term in the series:

$$H_0(y) = c_0$$
$$\psi_0(y) = Ac_0 e^{-\frac{y^2}{2}}. \tag{5.239}$$

Now, in order to go back to the x coordinate, we use $y = \sqrt{a}\, x = x\sqrt{\frac{m\omega}{\hbar}}$ and eq. 5.239 gives the first eigenfunction:

$$\psi_0(y) = Ac_0 e^{-\frac{m\omega y^2}{2\hbar}} = A_0 e^{-\frac{m\omega y^2}{2\hbar}}. \tag{5.240}$$

If $N = 1$ we know that we now have to impose $c_1 = 0$ and the only choice is $n = 0$. Eq. 5.238 gives:

$$H_1(y) = c_1 y$$
$$\psi_1(y) = Ac_1 y e^{-\frac{y^2}{2}}$$
$$\psi_1(x) = Ac_1 x\sqrt{\frac{m\omega}{\hbar}}\, e^{-\frac{m\omega}{2\hbar}x^2} = A_1 x\sqrt{\frac{m\omega}{\hbar}}\, e^{-\frac{m\omega}{2\hbar}x^2} \tag{5.241}$$

For $N = 2$, following the same procedure we get:

$$H_2(y) = c_0(1 - 2y^2)$$

$$\psi_2(y) = Ac_0(1 - 2y^2)e^{-\frac{y^2}{2}} \tag{5.242}$$

$$\psi_2(x) = Ac_0\left(1 - 2\frac{m\omega}{\hbar}x^2\right)e^{-\frac{m\omega}{2\hbar}x^2} = A_2\left(1 - 2\frac{m\omega}{\hbar}x^2\right)e^{-\frac{m\omega}{2\hbar}x^2}$$

For $N = 3$, we have:

$$H_3(y) = c_1\left(y - \frac{2}{3}y^3\right)$$

$$\psi_3(y) = Ac_1\left(y - \frac{2}{3}y^3\right)e^{-\frac{y^2}{2}} \tag{5.243}$$

$$\psi_3(x) = Ac_1 x\sqrt{\frac{m\omega}{\hbar}}(1 - \frac{2}{3}x^2)e^{-\frac{m\omega}{2\hbar}x^2} = A_3 x\sqrt{\frac{m\omega}{\hbar}}(1 - \frac{2}{3}x^2)e^{-\frac{m\omega}{2\hbar}x^2}$$

where the constant A from eq. 5.225 is incorporated into the new constants A_i. We can keep going and generate all the eigenfunctions $\psi_i(x)$. If we normalize[24] all the eigenfunctions we find that, written in a compact way, the general formula is:

$$\psi_i(x) = \left(\frac{m\omega}{\pi\hbar}\right)^{\frac{1}{4}} \frac{1}{\sqrt{2^i i!}} H_i(\sqrt{a}\,x)e^{-\frac{ax^2}{2}}. \tag{5.244}$$

---

[24]For details about the normalization of the eigenfunctions see the book from Schiff [38].

6 Matrices in Quantum Mechanics

We have seen in the previous chapters that classical mechanics was not able to explain a range of phenomena. The starting point was the inability to model mathematically the spectral behavior of the blackbody radiation. Planck started the quantum revolution with his quantization of the energy exchanged between the e.m. field and the walls of the blackbody. Einstein then extended the quantization to light itself with the photon hypothesis with the explanation of the photoelectric effect. Together with experiments on Compton's scattering it emerged that photons had both wave and particle behavior depending on the specific experiment.

Bohr used Planck's ideas to explain the line emission of hydrogen atoms which gave an indication that the newborn quantum mechanics was giving a better description of experiments. De Broglie, in an attempt to better understand the dynamics of the atom made the assumption that not only photons but matter in general had the duality behavior: pilot waves needed to be associated to massive particles to explain a range of experiments showing interference and diffraction. The main characteristic required to these pilot waves was the ability to interfere with themselves.

Then Schrödinger wrote his equation for this waves, now called *wavefunctions*, which represented a successful attempt at describing mathematically all the phenomena not described by classical mechanics. A new class of mathematical objects, procedures and concepts was introduced like, for example, wavefunctions, operators, expectation values, commutators, uncertainty principle, etc. In Schrödinger picture the operators are constant and the wavefunctions carry the time dependence.

Slightly before Schrödinger, Werner Heisenberg came up with a completely different mathematical formulation of quantum mechanics based on operators expressed as hermitian matrices. Few years later it was shown that the two formulations are equivalent and it is a matter of taste – or better convenience – which mathematics is used. In fact, certain problems are handled more easily in one formulation rather than the other. In Heisenberg picture, the wavefunctions are constant and the operators carry the time dependence.

Dirac proposed a sort of combination of Schrödinger and Heisenberg formulations called *interaction picture* particularly useful when the physical problems contains interactions that change with time both the wavefunctions and the operators.

Richard Feynman came up with another mathematical formulation of quantum mechanics: path integrals that we briefly introduced in the previous chapter. In this formulation the classical action principle is generalized to quantum mechanics. It was later demonstrated that Feynman approach is also equivalent to the other formulations.

DOI: 10.1201/9781003145561-6

In this chapter, we will *not* follow a rigorous mathematical description. We will rather use a "physicist" approach where mathematical assumptions are guided by physical intuition and mathematical proofs are not always rigorous. The interested reader can consult many books where the mathematical foundations of quantum mechanics are treated with the necessary accuracy.

6.1 DIRAC'S NOTATION

Let us introduce a very practical notation first proposed by P.A.M Dirac[12]. We have already mentioned that the state of a quantum system, i.e. a system governed by quantum mechanics, is completely determined if we know the wavefunction $\psi(x,t)$. In Dirac's notation, this state is called a *state vector* and is indicated by the symbol $|\psi\rangle$, also called **ket** . With this new notation, Schrödinger equation is written as:

$$i\hbar\frac{\partial}{\partial t}|\psi\rangle = \hat{H}|\psi\rangle. \tag{6.1}$$

Kets are subject to the same algebra of the more familiar vectors with the difference that they can be complex. In fact, given two complex numbers c_1 and c_2, starting from any two kets $|A\rangle$ and $|B\rangle$ we can always build a new ket such that:

$$|C\rangle = c_1|A\rangle + c_2|B\rangle. \tag{6.2}$$

Eq. 6.2, although apparently stating the obvious, is in reality a very powerful statement. This equation is telling us that, if the two states $|A\rangle$ and $|B\rangle$ are solutions of the same Schrödinger equation, then also $|C\rangle$ is a solution. A system described by the ket $|C\rangle$ is said to be in a state of **superposition** between the states $|A\rangle$ and $|B\rangle$.

Now suppose we build the following state $c_1|A\rangle + c_2|A\rangle$, i.e. a state formed by superposing it with itself via two nonzero coefficients c_1 and c_2. We have that:

$$c_1|A\rangle + c_2|A\rangle = (c_1 + c_2)|A\rangle. \tag{6.3}$$

Apart for the case $(c_1 + c_2) = 0$, for which the result of the superposition returns a null state, we make the assumption that the new state $(c_1 + c_2)|A\rangle$ is representing exactly the original state $|A\rangle$. This means that all kets $a|A\rangle$, i.e. kets multiplied by any complex number $a \neq 0$, represent exactly the same state. In terms of vectors, it means that what is important is the *direction* of the vector and not its magnitude.

Given the freedom of multiplying any ket without altering the state, it follows that in the ket $|C\rangle$ of eq. 6.2 only the ratio of the two coefficients c_1 and c_2 is needed. As a consequence, any state is determined by one complex number or, equivalently, two real numbers.

Having defined the ket $|\psi\rangle$, we now introduce a dual vector, called **bra** vector and indicated with the symbol $\langle\phi|$ such that the inner product is indicated with the symbol $\langle B|A\rangle$, i.e.:

$$\langle\phi|\psi\rangle = \int_{-\infty}^{+\infty} \phi^*(x)\psi(x)dx \tag{6.4}$$

where $\phi^*(x)$ is the complex conjugate of $\phi(x)$. It follows that, if we want to calculate the inner product of two ket vectors $|A\rangle$ and $|B\rangle$, we have to take the complex conjugate of one of the two, for example $|B\rangle$: $\langle B|A\rangle$. It can be shown that:

$$\langle B|A\rangle = \langle A|B\rangle^* \tag{6.5}$$

from which it follows that $\langle A|A\rangle \geq 0$ and is real. The normalization condition 5.58 for $\langle \psi|\psi\rangle$ is then written as:

$$\langle \psi|\psi\rangle = 1. \tag{6.6}$$

We have implicitly introduced Dirac's ket as a generalization of the concept of vector. In classical mechanics we think about a vector $\vec{v}$ as a physical quantity defined in a space – usually the tridimensional space (x, y, z) – and represented in terms of the base vectors $\vec{v} = v_x \vec{e}_x + v_y \vec{e}_y + v_z \vec{e}_z$, where $\vec{e}_x, \vec{e}_y, \vec{e}_z$ are unit vectors forming an orthogonal reference frame and v_x, v_y, v_z are the components of the vector $\vec{v}$ in such a reference frame. In complete analogy, we can think of any wavefunction as a vector in a space formed by an orthogonal set of wavefunctions called base state vectors $|\psi_i\rangle$:

$$|\psi\rangle = a_1 |\psi_1\rangle + a_2 |\psi_2\rangle + \ldots = \sum_i a_i |\psi_i\rangle. \tag{6.7}$$

This specific property is a direct consequence of one of the postulates of quantum mechanics.

6.2 LINEAR ALGEBRA

In the previous parts of this book we have introduced the wavefunctions as objects that, once completely specified, determine the status of a quantum system. In order to specify its time evolution, we then introduced Schrödinger equation which, when dealing with conservative systems, i.e. systems where the energy is conserved, reduces to the TISE – Time Independent Schrödinger equation.

In preparation to studying an alternative formulation of quantum mechanics due to Werner Heisenberg[1], in this section we introduce the concept of *linear vector space*.

In mathematics a space is a set of objects (also called "elements") subject to some relationships between them. Having introduced Dirac notation in section 6.1, let us identify the elements as vectors with an index j such that our space is formed by $|1\rangle$, $|2\rangle$, ..., $|j\rangle$, ... objects. We will restrict ourselves to the (brief) introductory study of linear vector spaces, where the elements are called *vectors* and where we define three operations:

1. Multiplication between vectors and scalars $a|1\rangle$
2. Addition between vectors $|1\rangle + |2\rangle$ $\tag{6.8}$
3. Multiplication between vectors $\langle 2|1\rangle$

---

[1]Werner Karl Heisenberg (1901-1976) was a German physicist considered one of the fathers of quantum mechanics. In 1925, with a series of groundbreaking papers, together with Max Born and Pascual Jordan, he introduced and developed the matrix formulation of quantum mechanics. He was also the first to formulate the uncertainty principle that goes under his name.

where a is a scalar. Notice that the the scalars can be complex numbers: in this case the space is called *complex vector space*[2].

Rule 1 has the following properties:

$$1a. \text{ Distributive in scalars, i.e. } (a+b)|1\rangle = a|1\rangle + b|1\rangle$$
$$1b. \text{ Distributive in vectors, i.e. } a(|1\rangle + |2\rangle) = a|1\rangle + a|2\rangle \qquad (6.9)$$
$$1c. \text{ Associative, i.e. } a(b|1\rangle) = (ab)|1\rangle$$

Rule 2 ha the following properties:

$$2a. \text{ Commutative, i.e. } |1\rangle + |2\rangle = |2\rangle + |1\rangle$$
$$2b. \text{ Associative, i.e. } |1\rangle + (|2\rangle + |3\rangle) = (|1\rangle + |2\rangle) + |3\rangle \qquad (6.10)$$

Let us add a few more properties:

$$3a. \text{ Null vector: exists a unique } |0\rangle \text{ such that } |1\rangle + |0\rangle = |1\rangle.$$
$$3b. \text{ Inverse vector: exists a unique } |-1\rangle \text{ such that } |1\rangle + |-1\rangle = |0\rangle \qquad (6.11)$$

It is very important to underline the fact that we call "vectors" the elements of the linear space without any connection with the vector concept in classical mechanics. The vectors of the linear space can be objects that are not even remotely connected to the classical vectors. The set of all the 2×2 matrices $\sigma = \begin{pmatrix} \sigma_{11} & \sigma_{12} \\ \sigma_{21} & \sigma_{22} \end{pmatrix}$, where the σ_{ij} are complex numbers, is a typical example of a linear vector space where the elements are not vectors in the classical way. It can be shown that they satisfy all the properties listed above.:

$$\alpha \cdot \begin{pmatrix} a & b \\ c & d \end{pmatrix} = \begin{pmatrix} \alpha a & \alpha b \\ \alpha c & \alpha d \end{pmatrix}$$

$$\alpha \cdot \left[\begin{pmatrix} a & b \\ c & d \end{pmatrix} + \begin{pmatrix} e & f \\ g & h \end{pmatrix} \right] = \begin{pmatrix} \alpha a & \alpha b \\ \alpha c & \alpha d \end{pmatrix} + \begin{pmatrix} \alpha e & \alpha f \\ \alpha g & \alpha h \end{pmatrix} = \alpha \cdot \begin{pmatrix} a & b \\ c & d \end{pmatrix} + \alpha \cdot \begin{pmatrix} e & f \\ g & h \end{pmatrix}$$

$$\alpha \left(\beta \begin{pmatrix} a & b \\ c & d \end{pmatrix} \right) = (\alpha\beta) \begin{pmatrix} a & b \\ c & d \end{pmatrix}$$

$$\begin{pmatrix} a & b \\ c & d \end{pmatrix} + \begin{pmatrix} e & f \\ g & h \end{pmatrix} = \begin{pmatrix} e & f \\ g & h \end{pmatrix} + \begin{pmatrix} a & b \\ c & d \end{pmatrix}$$

$$\begin{pmatrix} a & b \\ c & d \end{pmatrix} + \left[\begin{pmatrix} e & f \\ g & h \end{pmatrix} + \begin{pmatrix} i & j \\ k & l \end{pmatrix} \right] = \left[\begin{pmatrix} a & b \\ c & d \end{pmatrix} + \begin{pmatrix} e & f \\ g & h \end{pmatrix} \right] + \begin{pmatrix} i & j \\ k & l \end{pmatrix}$$

$$(6.12)$$

where $\alpha, \beta, a, b, .., l$ and are all scalars[3].

---

[2]If we want to be a little bit more accurate: the set of scalars a are referred to as *the field* over which the vector space is defined. If the scalars are real numbers then the vector space is a real vector space, while if the scalars are complex numbers then the vector space is called complex vector space.

[3]Notice that complex numbers are scalars.

Let's consider a set of vectors $|1\rangle, |2\rangle, ..., |N\rangle$. We say that the vectors are *linearly independent* if the following expression:

$$\sum_{i=1}^{N} = c_i |i\rangle = |0\rangle \tag{6.13}$$

is valid if and only if all the $c_i = 0$. If, on the other hand, it is possible to find a set of coefficients $c_i \neq 0$ for which eq. 6.13 is valid, then the set of vectors is called *linearly dependent*.

An important consequence of eq. 6.13 consists in the impossibility of writing any vector of the linearly independent set as a linear combination of the rest of the vectors. This can be easily seen if we consider the very simple case of $N = 3$ where the condition 6.13 is:

$$c_1 |1\rangle + c_2 |2\rangle + c_3 |3\rangle = 0 \tag{6.14}$$

Now, let's suppose that we can write the ket $|1\rangle$ as a linear combination of the other two kets $|2\rangle$ and $|3\rangle$:

$$|1\rangle = a |2\rangle + b |3\rangle \tag{6.15}$$

which can be rearranged as:

$$a |2\rangle + b |3\rangle - |1\rangle = 0. \tag{6.16}$$

Comparing eq. 6.16 with eq. 6.13 tells us that it is not possible to have the co-efficients of the ket $|1\rangle \neq 0$ in eq. 6.16 since the condition of linearly independence requires that all the coefficients in eq. 6.16 must be zero. Therefore the expression 6.15 is not allowed.

The condition of linear independence has a few important consequences. A set of vector is called a *basis* if its elements are linearly independent and if any element of the vector space can be expressed as a linear combination of the basis vectors. Let's assume that $|u\rangle$ is one element of the linear vector space and that $|i\rangle$, $i = 1, ..., N$, is a basis. Then we can always write for each vector belonging to the vector space, the following:

$$|u\rangle = \sum_{i=1}^{N} c_i |i\rangle \tag{6.17}$$

where the coefficients c_i are called *components* or *coordinates*.

A vector space can have many different basis. However, all basis have the same number of elements called *dimension* of the vector space.

The *span* of a vector space is another useful concept. The span of a set of vectors is the linear space formed by the set of all the vectors that can be written as linear combinations of the vectors belonging to the given set. For example, if the vector space contain just one element, then the span is the set of all the vectors multiplied by a scalar and the vector is also a basis. In this case we say that the basis vector *spans* the space.

It is instructive to consider classical geometrical vectors for this next example. Suppose that our linear vector space is made of just one 2-D vector $|u\rangle = \begin{pmatrix} 1 \\ 1 \end{pmatrix}$ (see

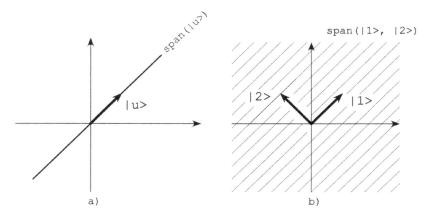

Figure 6.1 a) The span of the vector $|u\rangle$ is the straight line containing all vectors $a|u\rangle$ where a is a constant. b) The span of the two vectors $|1\rangle, |2\rangle$ is the shaded area, i.e. the plane passing through the two vectors.

fig. 6.1 left panel a)). Span($|u\rangle$) is the set of all linear combinations of $|u\rangle$ which is the straight line passing exactly through the vector $|u\rangle$.

If we now have two linearly independent geometrical vectors $|1\rangle = \begin{pmatrix} 1 \\ 1 \end{pmatrix}$ and $|2\rangle = \begin{pmatrix} -1 \\ 1 \end{pmatrix}$ the span($|1\rangle, |2\rangle$) is the plane passing through the two vectors $|1\rangle$ and $|2\rangle$ (the shaded area in fig. 6.1 right panel b)). It is clear that, if the two vectors were aligned, then the span would collapse to a line but that would mean that the two vectors were not linearly independent since one is a multiple of the other.

6.2.1 PAULI MATRICES

Let's go back to our linear vector space of 2×2 matrices and define the following 4 matrices[4]:

$$\sigma_1 = \begin{pmatrix} 0 & 1 \\ 1 & 0 \end{pmatrix}$$

$$\sigma_2 = \begin{pmatrix} 0 & -i \\ i & 0 \end{pmatrix}$$

$$\sigma_3 = \begin{pmatrix} 1 & 0 \\ 0 & -1 \end{pmatrix} \tag{6.18}$$

$$\sigma_4 = \begin{pmatrix} 1 & 0 \\ 0 & 1 \end{pmatrix}$$

---

[4]The first three matrices are called *Pauli matrices* and are used in Pauli's equation describing spin 1/2 particles.

and verify that they are a basis. This is achieved by requesting that any 2×2 matrix can be expressed as a linear combination of the Pauli matrices. If $\begin{pmatrix} a & b \\ c & d \end{pmatrix}$ is a generic matrix 2×2 matrix, we have:

$$\begin{pmatrix} a & b \\ c & d \end{pmatrix} = c_1 \begin{pmatrix} 0 & 1 \\ 1 & 0 \end{pmatrix} + c_2 \begin{pmatrix} 0 & -i \\ i & 0 \end{pmatrix} + c_3 \begin{pmatrix} 1 & 0 \\ 0 & -1 \end{pmatrix} + c_4 \begin{pmatrix} 1 & 0 \\ 0 & 1 \end{pmatrix}. \qquad (6.19)$$

Where we have substituted the Pauli matrices of eq. 6.18 into eq. 6.19. We have:

$$\begin{pmatrix} a & b \\ c & d \end{pmatrix} = \begin{pmatrix} c_3 + c_4 & c_1 - ic_2 \\ c_1 + ic_2 & -c_3 + c_4 \end{pmatrix}. \qquad (6.20)$$

If we recall that two matrices are equal if and only if the elements are equal, then the coefficients of the expansion 6.19 are found by solving the four equations in four unknown:

$$\begin{cases} a & = c_3 + c_4 \\ b & = c_1 - ic_2 \\ c & = c_1 + ic_2 \\ d & = -c_3 + c_4 \end{cases} \qquad (6.21)$$

which proves that we can always express any 2×2 matrix in terms of the Pauli matrices. Therefore Pauli matrices are a basis for the linear vector space of 2×2 matrices. Notice also the important consequence of eq. 6.21: we have *exactly* four equations for four unknown meaning that the number of basis needed to span the space of 2×2 matrices is exactly four.

We can generalize the previous result by stating that any vector $|v\rangle$ in an N-dimensional space can be written as a linear combination of exactly N linearly independent vectors[5]. The choice of basis vector is clearly arbitrary and there are infinite basis vectors that we can choose.

We are left to discuss the definition of *length* of a vector. While the concept of "length" of a vector is quite obvious in the case of geometrical vectors, in our generalized pictures the elements of the space, there exist objects that do not have an immediate definition of length. The case of 2×2 matrices that we have previously discussed is just one example. What is the "length" of a matrix?

We can attempt to assign a length to the elements of the vector space by defining a **dot product**. In complete analogy to the case of geometrical vectors for which, given two vectors $\vec{u} = (u_1, u_2, ..., u_N)$ and $\vec{v} = (v_1, v_2, ..., v_N)$ their dot product is simply $\vec{u} \cdot \vec{v} = u_1 v_1 + u_2 v_2 + ... + u_N v_N$, we define the dot product of two elements of the linear vector space $|u\rangle$ and $|v\rangle$, we call it **the inner product** and we denote it by the symbol $\langle u|v\rangle$.

---

[5]This statement can be easily proved by supposing that there exists one vector $|u\rangle$ that cannot be expressed as a linear combination of the basis vectors. But then we could add this vector to the set of basis vectors increasing by one unit the number of basis vectors. This would mean that now we have N+1 basis vectors in an N-dimensional space which we have seen is not possible.

We also impose that our newly defined inner product obeys the following axioms:

$$\langle u|v \rangle = \langle v|u \rangle^*$$
$$\langle u|u \rangle \geq 0 \qquad (6.22)$$
$$\langle u| (a|v \rangle + b|w \rangle) = a\langle u|v \rangle + b\langle u|w \rangle$$

where the asterisk symbol means complex conjugation and we use the convenient notation $|u+v \rangle = |u \rangle + |v \rangle$ and $|au \rangle = a|u \rangle$ where a is a constant. Once we have defined an inner product then our linear vector space over the field of complex number is called an **Hilbert space**.

Let's give another important definition: two vectors are called **orthogonal** if their inner product is equal to zero. It follows that a basis is called orthogonal if its base vectors are orthogonal to each other. In addition, if the inner product of each basis vector with itself is unity, then the basis is called **orthonormal**. If we indicate an orthonormal basis with the symbol $|e_i \rangle$, with $i = 1, 2, ..., N$, then the basis satisfies:

$$\langle e_i|e_j \rangle = \delta_{ij} \qquad (6.23)$$

where δ_{ij} is the Kronecker symbol already defined in eq. 5.158.

We can think of an orthonormal basis as a basis of unit vectors giving the unit of measure to all the vectors spanned. In fact, a very important feature of orthonormal basis is that, for each vector $|u \rangle$ belonging to the vector space, the coefficients c_i of the expansion:

$$|u \rangle = \sum_{i=1}^{N} c_i |e_i \rangle \qquad (6.24)$$

are given by the inner product of the vector $|u \rangle$ with the basis vectors according to:

$$c_i = \langle e_i|u \rangle \qquad (6.25)$$

which is equivalent to the projection of a geometrical vector onto its orthogonal unit axis.

We now go back to the Pauli matrices and verify that they are not only a basis but they are orthonormal with respect to an inner product defined by:

$$\langle u|v \rangle = \frac{1}{2} \text{Tr}(u^\dagger v) \qquad (6.26)$$

where u and v are two 2×2 matrices and the operation $Tr(u)$, called *the trace*, is the sum of the diagonal elements of the matrix:

$$Tr(u) = \sum_{i=1}^{N} u_{ii} \qquad (6.27)$$

The inner product 6.26 is called *Hilbert-Schmidt* inner product.

Before verifying that the inner product 6.26 is a valid inner product, i.e. it satisfies the axioms 6.22, let's define the conjugate transposed of a matrix u written, as in the case of operators of previous chapters, with the symbol $u^\dagger$.

The conjugate transposed of the matrix u is the matrix obtained by transposing it and then take the complex conjugate of its elements. If $u = \begin{pmatrix} a & b \\ c & d \end{pmatrix}$ then its conjugate transposed is:

$$u^\dagger = \begin{pmatrix} a^* & c^* \\ b^* & d^* \end{pmatrix} \tag{6.28}$$

or in terms of its components:

$$u_{ij}^\dagger = u_{ji}^*. \tag{6.29}$$

Notice that, in analogy with the differential operators discussed in the previous chapter, a matrix is called **Hermitian** if it is equal to its conjugate transposed, i.e. $u = u^\dagger$.

Going back to the inner product, it follows that for any matrix u we have:

$$Tr(u^\dagger) = (Tr(u))^*. \tag{6.30}$$

It follows that:

$$\langle u|v \rangle = \frac{1}{2}Tr(u^\dagger v) = \frac{1}{2}(Tr(v^\dagger u))^\dagger = \frac{1}{2}Tr(v^\dagger u)^* = (\langle v|u \rangle)^* \tag{6.31}$$

where we used the rule:

$$(uv)^\dagger = v^\dagger u^\dagger \tag{6.32}$$

which can be proved by writing the rule for the product of two matrices u and v whose components are:

$$(uv)_{ij} = \sum_{i=1}^{N} u_{ij} v_{jk} \tag{6.33}$$

and writing the components of the transpose of the product, indicated with the symbol T:

$$(uv)_{ij}^T = (uv)_{ji} = \sum_{k=1}^{N} u_{ik} v_{kj} = \sum_{k=1}^{N} (v^T)_{jk} (u^T)_{ki}. \tag{6.34}$$

The complex conjugate of eq. 6.34 proves eq. 6.32 and thus eq. 6.31 thus proves axiom 1.

We now proceed to axiom 2 in 6.22. Let's calculate the inner product of $|u\rangle$ with itself. Using the previous results, we have:

$$\langle u|u \rangle = \frac{1}{2}Tr(u^\dagger u) = \sum_{i=1}^{N}\sum_{j=1}^{N} (u^\dagger)_{ij} u_{ji} = \sum_{i=1}^{N}\sum_{j=1}^{N} (u^*)_{ji} u_{ji} = \sum_{i=1}^{N}\sum_{j=1}^{N} |u_{ji}|^2 \geq 0 \tag{6.35}$$

which proves the second axiom. Finally, in order to prove the third axiom we divide the proof into two separate parts. First we prove that $\langle u|av \rangle = a\langle u|v \rangle$ and then we prove that $\langle u|v+w \rangle = \langle u|v \rangle + \langle u|w \rangle$.

Let's start by noticing that the trace operation is linear, i.e.:

$$Tr(au + bv) = aTr(u) + bTr(u). \tag{6.36}$$

It follows that:

$$\langle u|av \rangle = \frac{1}{2}\text{Tr}\left(u^\dagger(av)\right) = \frac{1}{2}\text{Tr}\left(au^\dagger v\right) = \frac{1}{2}a\,\text{Tr}\left(u^\dagger v\right) = a\langle u|v \rangle. \tag{6.37}$$

Finally:

$$\langle u|v+w \rangle = \frac{1}{2}Tr(u^\dagger(v+w)) = \frac{1}{2}Tr(u^\dagger v) + \frac{1}{2}Tr(u^\dagger w) = \langle u|v \rangle + \langle u|w \rangle \tag{6.38}$$

which complete the proof that the proposed inner product satisfies the three axioms 6.22.

In addition to being hermitian, Pauli matrices are also involutory and unitary. A matrix u that is also its own inverse is called *involutory*. In general, the inverse of a matrix u is indicated with the symbol u^{-1} and is such that $uu^{-1} = I$ where the matrix I is the identity matrix defined as $I = \begin{pmatrix} 1 & 0 \\ 0 & 1 \end{pmatrix}$. It follows that all the involutory matrices satisfy $uu^{-1} = uu = u^2 = I$, i.e. the square of an involutory matrix is the identity. As a consequence, an involutory matrix is the square root of the identity matrix. According to this definition, the identity matrix is also involutory. It is easy to show that Pauli matrices are involutory:

$$\sigma_1 \cdot \sigma_1 = \begin{pmatrix} 0 & 1 \\ 1 & 0 \end{pmatrix} \cdot \begin{pmatrix} 0 & 1 \\ 1 & 0 \end{pmatrix} = \begin{pmatrix} 0 & 1 \\ 1 & 0 \end{pmatrix}$$

$$\sigma_2 \cdot \sigma_2 = \begin{pmatrix} 0 & -i \\ i & 0 \end{pmatrix} \cdot \begin{pmatrix} 0 & -i \\ i & 0 \end{pmatrix} = \begin{pmatrix} 0 & 1 \\ 1 & 0 \end{pmatrix} \tag{6.39}$$

$$\sigma_3 \cdot \sigma_3 = \begin{pmatrix} 1 & 0 \\ 0 & -1 \end{pmatrix} \cdot \begin{pmatrix} 1 & 0 \\ 0 & -1 \end{pmatrix} = \begin{pmatrix} 0 & 1 \\ 1 & 0 \end{pmatrix}$$

It follows that the inner product of Pauli matrices with themselves is equal to 1:

$$\langle \sigma_i | \sigma_i \rangle = \frac{1}{2}Tr(\sigma_i \cdot \sigma_i) = \frac{1}{2}Tr(I) = 1 \tag{6.40}$$

which shows that all Pauli matrices are normalized to 1 under the inner product 6.26. We see also *a posteriori* why the inner product contains the factor $\frac{1}{2}$.

Finally, we show that Pauli matrices are orthogonal and, since they are normalized, they are orthonormal.

In order to do so, let's study the product of two Pauli matrices. We have seen already that the product of a Pauli matrix with itself is equal to the identity matrix. It can be shown by direct calculation that the product of any two Pauli matrices is equal to another Pauli matrix. An exhaustive calculation will show that we can express the product of two Pauli matrices by:

$$\sigma_i \sigma_j = I\delta_{ij} + i\varepsilon_{ijk}\sigma_k \tag{6.41}$$

where ε_{ijk} is the Levi-Civita symbol defined by:

$$\varepsilon_{i,j,k} = \begin{cases} 0 & \text{for } i = j, j = k, k = i \\ +1 & \text{for } (i,j,k) \text{ any combination of } (1,2,3),(2,3,1),(3,1,2) \\ -1 & \text{for } (i,j,k) \text{ any combination of } (1,23,2),(3,2,1),(2,1,3). \end{cases} \tag{6.42}$$

Let's write the inner product of two different Pauli matrices. From eq. 6.41, we have that $\sigma_i \sigma_j = i\varepsilon_{ijk}\sigma_k$, for $i \neq k$. Therefore:

$$\langle \sigma_i | \sigma_j \rangle = \frac{1}{2}Tr(\sigma_i \cdot \sigma_j) = \frac{i}{2}\varepsilon_{ijk}Tr(\sigma_k) = 0. \tag{6.43}$$

The last equality is justified by noticing that Pauli matrices have all zero trace.

6.2.2 MATRICES IN QUANTUM MECHANICS

In the previous section, we introduced the abstract concept of linear vector space as a set of elements, called vectors, which are subject to a set of relationships like 6.8. In order to point out that the concept of "vector" is abstract and not connected in principle with the concept of geometric vector, we showed that we can build a legitimate linear vector space whose elements are the set of 2×2 matrices. We have seen that the set of Pauli matrices plus the identity matrix form an orthonormal basis with respect to the Hilbert-Schmidt inner product.

We now leave this space, although we will use Pauli matrices later. In this section we make the connection with physics and show that vectors in Hilbert space represent the quantum states of the system. In this case, the vectors are constructed in such a way that their norm $\langle \psi | \psi \rangle = 1$ in order to properly represent a probability. A generic wavefunction is the inner product $\langle x | \psi \rangle$ with the ket $|x\rangle$ representing the space coordinate x.

One of the simplest (perhaps!) quantum states that we can imagine in a 2-dimensional Hilbert space, where the system can assume only two possible values. In other words, if we make a measurement on such a system we obtain either of two eigenvalues with associated two eigenfunctions that, for simplicity, we write as $|0\rangle$ and $|1\rangle$.

Let's introduce a different Hilbert space where the vectors are two-dimensional column vectors. An example of ket vectors would be:

$$|0\rangle = \begin{pmatrix} 1 \\ 0 \end{pmatrix}, \quad |1\rangle = \begin{pmatrix} 0 \\ 1 \end{pmatrix} \tag{6.44}$$

while an example of bra vectors would be:

$$\langle 0| = \begin{pmatrix} 1 & 0 \end{pmatrix}, \quad \langle 1| = \begin{pmatrix} 0 & 1 \end{pmatrix} \tag{6.45}$$

A particular Hilbert space where the above vectors live would be composed by all 2-elements column vectors with the inner product defined as the standard vector multiplication. If $|a\rangle = \begin{pmatrix} a_1 & a_2 \end{pmatrix}$ and $|b\rangle = \begin{pmatrix} b_1 & b_2 \end{pmatrix}$ then $\langle a|b \rangle = a_1 b_1 + a_2 b_2$ which is the standard dot product[6].

---

[6]It can be shown that the dot product satisfies the axioms 6.22 and therefore the space we are considering is an Hilbert space.

In complete analogy with differential operators acting on wavefunctions and generating another wavefunction, in our two-dimensional Hilbert space the 2×2 matrices represents operators that acting on a vector generate another vector. From now on we will indicate matrices, i.e. operators, with capital letters. Therefore a matrix A operating on the vector $|a\rangle$ to produce the vector $|b\rangle$ is indicated with $|b\rangle = A|a\rangle$.

It is easy to verify that the two vectors 6.44 form an orthonormal basis that we indicate as $|e_j\rangle$ with $|e_0\rangle = |0\rangle$ and $|e_1\rangle = |1\rangle$. In this case, the matrix elements of A can be written as:

$$A_{ij} = \langle e_i|A|e_j\rangle \tag{6.46}$$

which is telling us that the elements of a matrix are interpreted as the projection on the orthonormal axis.

We have seen that quantum mechanics needs to be expressed using complex numbers. Complex conjugation is an important operation for complex numbers and we know that is defined as follows: given a complex number $a = a + ib$ where a and b are real numbers, its complex conjugate is indicated with $a^* = a - ib$. Is there an equivalent operation for vectors and matrices? The answer is yes: the equivalent of a complex conjugate of a ket vector is a bra vector and the equivalent of a complex conjugate of a matrix is called adjoint or hermitian conjugate. We see here another analogy with differential operators.

Given the vectors $|\phi\rangle$ and $|\psi\rangle$, the hermitian conjugate of the matrix A is defined by:

$$(\langle\phi|A|\psi\rangle)^* = \langle\psi|A^\dagger|\phi\rangle . \tag{6.47}$$

Expressed in an orthonormal basis $|e_i\rangle$, by using 6.46 and 6.47, we have a recipe for calculating the hermitian conjugate of a matrix:

$$(\langle e_i|A|a_j\rangle)^* = A_{ij}^* = \langle e_j|A^\dagger|e_i\rangle = A_{ji}^* \tag{6.48}$$

which means that the hermitian conjugate of a matrix A is calculated by transposing the matrix and then taking the complex conjugate of its components. The adjoint operation has the following properties:

$$(cA)^\dagger = a^* A^\dagger$$
$$(AB)^\dagger = B^\dagger A^\dagger \tag{6.49}$$
$$|\psi\rangle^\dagger = \langle\psi| .$$

For each matrix A it is possible to calculate its *eigenvalues* λ_i and its *eigenvectors* $|u_i\rangle$ defined by the eigenvalue equation:

$$A|u_i\rangle = \lambda_i|u_i\rangle . \tag{6.50}$$

It is interesting to calculate the matrix A when expressed in the basis of its own eigenvectors. First, eq. 6.50 is in reality a set of N equations, where N is the dimension of the vector space. Then:

$$A|u_1\rangle = \lambda_1 |u_1\rangle$$
$$A|u_2\rangle = \lambda_2 |u_2\rangle$$
$$\vdots$$
$$A|u_N\rangle = \lambda_N |u_N\rangle$$

$$(6.51)$$

and eq. 6.51 can be re-written as:

$$A \begin{pmatrix} u_1 \\ u_2 \\ \vdots \\ u_N \end{pmatrix} = D \begin{pmatrix} u_1 \\ u_2 \\ \vdots \\ u_N \end{pmatrix} = \begin{pmatrix} \lambda_1 & 0 & 0 & 0 \\ 0 & \lambda_2 & 0 & 0 \\ 0 & 0 & \ddots & 0 \\ 0 & 0 & 0 & \lambda_N \end{pmatrix} \begin{pmatrix} u_1 \\ u_2 \\ \vdots \\ u_N \end{pmatrix} \qquad (6.52)$$

Eq. 6.52 shows that a matrix A, when expressed into the basis of its own eigenvectors, becomes diagonal and the elements on the diagonal are its eigenvalues. The matrix D is zero everywhere but on the diagonal. The matrices A and D share the same fundamental properties because they are related through a change of basis and are said to be *similar*. In other words, two (square) matrices are said to be similar if they represent the same linear operator expressed in different bases. It follows that two similar matrices have the same trace, rank[7], determinant and eigenvalues. It follows that the trace of a matrix A is the sum of its eigenvalues while the determinant is the product of its eigenvalues. Therefore, given a square matrix A, we have:

$$\text{Tr}\{A\} = \sum_i \lambda_i$$
$$\det(A) = \prod_i \lambda_i. \qquad (6.53)$$

There is an alternative, although equivalent, definition of similarity: two $N \times N$ matrices A and B are said to be similar if it exists an *invertible* matrix P such that:

$$B = P^{-1}AP. \qquad (6.54)$$

The matrix P is called *change of basis* matrix and must have a non zero determinant. The determinant of a square matrix A, indicated with the symbol $\det\{A\}$, is a number calculated from the matrix elements. The determinant is nonzero if and only if the matrix is invertible, i.e. given a matrix A with nonzero determinant it exists, and is unique, a matrix A^{-1} such that $AA^{-1} = A^{-1}A = I$, where I is the identity matrix.

The inverse of a square matrix is a long and tedious calculation. Given a matrix A, the general formula is:

$$A^{-1} = \frac{1}{\det\{A\}} A^C \qquad (6.55)$$

---

[7]The rank of a matrix is the dimension of the vector space generated by its columns. An equivalent definition: the rank fo a matrix is the number of linearly independent rows or columns.

where T^C is the matrix of cofactors. Each element A_{ij} of the matrix has a cofactor calculated by multiplying $(-1)^{i+j}$ times the determinant of the submatrix obtained by eliminating the ith row and the jth column.

The case of a 2×2 matrix is relatively simple. Given a 2×2 matrix $A = \begin{pmatrix} a & b \\ c & d \end{pmatrix}$ the inverse is given by:

$$A^{-1} = \frac{1}{\det\{A\}} \begin{pmatrix} d & -b \\ -c & a \end{pmatrix}. \tag{6.56}$$

The determinant of A is calculated as follows:

$$\det\{A\} = \begin{vmatrix} a & b \\ c & d \end{vmatrix} = ad - bc. \tag{6.57}$$

The determinant of a 3×3 matrix A is calculated as follows:

$$\det\{A\} = \begin{vmatrix} a & b & c \\ d & e & f \\ g & h & k \end{vmatrix} = a(ek - fh) - b(dk - fg) + c(dh - eg). \tag{6.58}$$

It is one of the fundamental postulates of quantum mechanics that to each observable is associated a matrix whose eigenvalues are the results of all possible measurements. Since eigenvalues are the results of measurements, they must be real numbers. This postulate can be expressed mathematically by requiring that the matrix is Hermitian, i.e. it is equal to its adjoint[8]. Looking at eq. 6.47 a Hermitian matrix A is a matrix for which:

$$A = A^\dagger \tag{6.59}$$

which in terms of components:

$$A_{ij} = A_{ji}^*. \tag{6.60}$$

Let's prove that the eigenvalues of a Hermitian matrix are real. Let A be a Hermitian matrix whose eigenvectors are denoted by $|u\rangle$ and eigenvalues λ. We have:

$$\begin{aligned} A|u\rangle &= \lambda|u\rangle \\ \langle u|A|u\rangle &= \lambda \langle u|u\rangle = \lambda \\ (\langle u|A|u\rangle)^\dagger &= \lambda^* \\ \langle u|A^\dagger|u\rangle &= \lambda^* \\ \langle u|A|u\rangle &= \lambda^* \\ \Rightarrow \lambda &= \lambda^* \end{aligned} \tag{6.61}$$

Let's now prove that the eigenvectors corresponding to distinct eigenvalues of a Hermitian matrix form an orthogonal basis. Let's assume the matrix A is Hermitian

---

[8] Sometimes Hermitian matrices are called self-adjoint.

and study the simple case of two eigenvalues. Let's call $|u\rangle$ the eigenvector corresponding to the eigenvalue λ_1 and let's call $|v\rangle$ the eigenvector corresponding to the eigenvalue λ_2. Therefore we have two eigenvalue equations:

$$A|u\rangle = \lambda_1 |u\rangle$$
$$A|v\rangle = \lambda_2 |v\rangle . \qquad (6.62)$$

If the eigenvalues are distinct, then we must have $(\lambda_1 - \lambda_2) \neq 0$. Let's multiply the first equation in 6.62 by $\langle v|$:

$$\langle v|A|u\rangle = \langle u|A^\dagger|v\rangle$$
$$= \langle u|A|v\rangle \qquad (6.63)$$
$$= \lambda_1 \langle v|u\rangle .$$

Let's now multiply the second equation in 6.62 by $\langle u|$:

$$\langle u|A|v\rangle = \lambda_2 \langle u|v\rangle$$
$$= \lambda_2^* \langle v|u\rangle \qquad (6.64)$$
$$= \lambda_2 \langle v|u\rangle .$$

Equations 6.63 and 6.64 clearly implies that $\lambda_1 \langle v|u\rangle = \lambda_2 \langle v|u\rangle$ which can be written as:

$$\lambda_1 \langle v|u\rangle = \lambda_2 \langle v|u\rangle$$
$$(\lambda_1 - \lambda_2) \langle v|u\rangle = 0 \qquad (6.65)$$
$$\Rightarrow \langle v|u\rangle = 0$$

where the last equality holds because $(\lambda_1 - \lambda_2) \neq 0$.

Although an $N \times N$ matrix has N eigenvalues, it might occur that not all are distinct. This means that to some specific eigenvalue corresponds more than one eigenvector. In this case the eigenvalue is said to be *degenerate*. We will come back to this issue later in this section when we discuss the diagonalization procedure.

Going back to the similarity transformation 6.54, in order to be sure that the change of basis return the operator in the same Hilbert space, we need to require that the inner product is invariant, i.e. it does not change when we operate a change of base. A unitary matrix therefore preserves the magnitude of all vectors. This is true if and only if:

$$A^\dagger A = AA^\dagger = I \qquad (6.66)$$

or, alternatively, $A^\dagger = A^{-1}$. In order to prove that unitary matrices preserve inner product, we start by noticing that, because of 6.49, given an orthonormal basis $|e_i\rangle$ we have that:

$$(A|e_i\rangle)^\dagger = \langle e_i|A^\dagger \qquad (6.67)$$

which gives us a rule to take the hermitian of a vector obtained when an operator A acts on a given vector. Eq. 6.67 can be interpreted as follows: if an operator A is acting on a ket, it is written on its left like, for example, $A|e_i\rangle$; the same operator, when operating on the bra, it is written on its right and is assumed to act toward its left, i.e. $\langle e_i|A^\dagger$. The Hermitian of a vector ket, interpreted as a column vector like eq. 6.44, is a vector bra interpreted as a row vector with the complex conjugate of its components. For example, if $|u\rangle = \begin{pmatrix} a \\ b \end{pmatrix}$, then its Hermitian conjugate is:

$$(|u\rangle)^\dagger = \langle u| = \begin{pmatrix} a^* & b^* \end{pmatrix}. \tag{6.68}$$

With this definition, the inner product can be written as $\langle u|u\rangle = a^*a + b^*b$. Let's verify that the inner product is preserved by unitary matrices. Given a matrix A acting on an orthonormal basis $|u\rangle = A|e_i\rangle$, with a corresponding $\langle v| = \langle e_j|A^\dagger$, we preserve the inner product if $\langle u|v\rangle = \langle e_j|e_i\rangle$. We have:

$$\langle u|v\rangle = \langle e_j|A^\dagger A|e_i\rangle = \langle e_j|e_i\rangle \text{ if and only if } A^\dagger A = I. \tag{6.69}$$

Unitary matrices have the property that their rows and columns form orthonormal sets.

Let's introduce a new operator called **Projection Operator** and defined by:

$$P_i = |e_i\rangle\langle e_i| \tag{6.70}$$

where $|e_i\rangle$ is an orthonormal basis. The operator P has the interesting property $P^2 = P$ and it "projects" any vector along the $|e_i\rangle$ axis[9]. In fact, if $|\psi\rangle = \sum_i c_i|e_i\rangle$ is a vector belonging to the space spanned by the basis $|e_i\rangle$, we have:

$$\begin{aligned} P_n|\psi\rangle &= P_n \cdot \sum_i c_i|e_i\rangle \\ &= |e_n\rangle\langle e_n| \cdot \sum_i c_i|e_i\rangle \\ &= \sum_i c_i \langle e_n|e_i\rangle|e_n\rangle \\ &= \sum_i c_i \delta_{ni}|e_n\rangle \\ &= c_n|e_n\rangle \end{aligned} \tag{6.71}$$

where we have used the orthonormality condition $\langle e_n|e_i\rangle = \delta_{ni}$.

Intuitively we would expect that if we add up all the projections of a vector along an orthonormal basis we obtain the same original vector. In fact, we have:

$$\begin{aligned} \sum_i P_i|\psi\rangle &= \sum_i |e_i\rangle\langle e_i| \left(\sum_j c_j|e_j\rangle \right) \\ &= \sum_{ij} c_j|e_i\rangle\langle e_i|e_j\rangle \\ &= \sum_i c_i|e_i\rangle \\ &= |\psi\rangle \end{aligned} \tag{6.72}$$

[9]The operator P projects a vector on a basis. Once it has been projected, repeated projections return the same vector thus qualitatively explaining the identity.

from which we write the **completeness relation**:

$$\sum_i |e_i\rangle \langle e_i| = I. \tag{6.73}$$

The projection operator allows us to write a diagonal matrix D in a simple form:

$$D = \sum_i |e_i\rangle \lambda_i \langle e_i|. \tag{6.74}$$

The basis $|e_i\rangle$ is not the only orthogonal basis in which we can express the matrix D. Let's suppose we have another orthogonal basis $|h_i\rangle$ and let's assume that the unitary matrix U changes basis from $|e_i\rangle$ to $|h_i\rangle$. In this case we can write that $|h_i\rangle = U|e_i\rangle$ and $\langle e_i|U^\dagger = \langle h_i|$.

We now ask, what happens if we change basis to the operator D of eq. 6.74? In order to evaluate the change of basis we use a simple identity $D = IDI$ where we use form 6.73 for the identity operator I. We have:

$$\begin{aligned}
D = IDI &= \sum_{ijk} |h_i\rangle \langle h_i|e_k\rangle \lambda_k \langle e_k|h_j\rangle \langle h_j| \\
&= \sum_{ijk} |h_i\rangle \langle e_i|U^\dagger|e_k\rangle \lambda_k \langle e_k|U|e_j\rangle \langle h_j| \\
&= \sum_{ij} |h_i\rangle \left(\sum_k U_{ik}^\dagger \lambda_k U_{kj} \right) \langle h_j| \\
&= \sum_{ij} |h_i\rangle B_{ij} \langle h_j|
\end{aligned} \tag{6.75}$$

where we have defined:

$$B = \sum_k U_{ik}^\dagger \lambda_k U_{kj}. \tag{6.76}$$

Eq. 6.76 can be written as:

$$B = U^\dagger D U. \tag{6.77}$$

Operators satisfying eq. 6.77 are called **normal**. The diagonal elements of D are the eigenvalues λ_k, also called **the spectrum** of B. The column vectors of U are such that $|b_j\rangle = \sum_i U_{ij}|e_j\rangle$. The column vectors of U are the normalized eigenvectors of B, i.e. $B|b_j\rangle = \lambda_j|b_j\rangle$.

Although the matrix D is diagonal, the matrix B is, in general, non diagonal. However, eq. 6.77 implies that, if the matrix U is unitary, then:

$$\begin{aligned}
B &= U^\dagger D U \\
UB &= UU^\dagger D U \\
UBU^\dagger &= DUU^\dagger \\
D &= UBU^\dagger
\end{aligned} \tag{6.78}$$

where we have used $UU^{\dagger} = I$. Eq. 6.78 shows that we can diagonalize the square matrix B by finding a unitary matrix U such that $D = UBU^{\dagger}$ where D is diagonal.

Any normal operator B admits the spectral decomposition given by:

$$B = \sum_i \lambda_i \langle b_i | b_i \rangle \qquad (6.79)$$

where $|b_i\rangle$ is an orthonormal basis. More in general it can be shown[10] that the eigenvectors of a Hermitian matrix A span the vector space where the matrix is defined. As a consequence, any vector within the space can be written in terms of these eigenvectors as:

$$|\psi\rangle = \sum_{i=1}^{N} c_i |u_i\rangle \qquad (6.80)$$

where the c_i are real coefficients and the $|u_i\rangle$ are the eigenvectors of the Hermitian matrix A.

Having realized its importance, we now describe the process of finding eigenvalues and eigenvectors of a matrix. We will restrict ourselves to Hermitian matrices because of their relationship with physical observables.

Given a Hermitian matrix A, the eigenvalue equation is:

$$A |a_i\rangle = \lambda_i |a_i\rangle \qquad (6.81)$$

where the ket $|a_i\rangle$ are the eigenvectors of the matrix A with eigenvalues λ_i. Eq. 6.81 corresponds to the search of those numbers (eigenvalues) for which the action of the operator A on the kets $|a_i\rangle$ results in a simple multiplication by a (different) number. We have seen previously that if the matrix is Hermitian than the eigenvalues are real numbers. This fact should not surprise us because we want to make sure that the spectrum of eigenvalues corresponds to the result of all the possible measurements which are obviously real numbers. For simplicity we consider here that there is only one independent eigenvector for each eigenvalue. The case of more than one eigenvector corresponding to the same eigenvalue will be considered later.

Let's now describe the procedure of finding the eigenvalues and the corresponding eigenvectors of a matrix A. We start by rewriting eq. 6.81 as:

$$(A - \lambda I) |a\rangle = 0 \qquad (6.82)$$

where the "0" on the right-hand side of eq. 6.82 is a vector with all zeros as components. In a more explicit form, eq. 6.82 can be written as:

$$\begin{pmatrix} A_{11} - \lambda_1 & A_{12} & \cdots & A_{1N} \\ A_{21} & A_{22} - \lambda_2 & \cdots & A_{2N} \\ \vdots & \vdots & \vdots & \vdots \\ A_{N1} & A_{N2} & \cdots & A_{NN} - \lambda_N \end{pmatrix} |a\rangle = 0. \qquad (6.83)$$

---

[10] For a proof see, for example, Byron and Fuller [7].

Eq. 6.83 is equivalent to a set of N linear homogeneous equations which has a non trivial solution, i.e. $|a\rangle = 0$, if the determinant of the matrix in eq. 6.83 is equal to zero:

$$\det(A - \lambda I) = \begin{vmatrix} A_{11} - \lambda_1 & A_{12} & \cdots & A_{1N} \\ A_{21} & A_{22} - \lambda_2 & \cdots & A_{2N} \\ \vdots & \vdots & \vdots & \vdots \\ A_{N1} & A_{N2} & \cdots & A_{NN} - \lambda_N \end{vmatrix} = 0 \qquad (6.84)$$

which, when calculated, is equivalent to the so called *characteristic equation*[11] consisting of a N-th degree polynomial $p(\lambda)$, in the unknown λ, equal to zero, i.e.:

$$p(\lambda) = 0 \qquad (6.85)$$

The solutions of eq. 6.85 are the eigenvalues which, when inserted back into eq. 6.81, determine the eigenvectors. From the theory of linear systems of equations, 6.85 has N distinct solutions corresponding to N linearly independent eigenvectors. The case of degenerate eigenvalues will be discussed later.

An example might clarify the process. Let's find the eigenvalues and the eigenvectors of Pauli matrices. For σ_1 we have:

$$\begin{pmatrix} 0 & 1 \\ 1 & 0 \end{pmatrix} \begin{pmatrix} u_1 \\ u_2 \end{pmatrix} = \lambda \begin{pmatrix} u_1 \\ u_2 \end{pmatrix}$$
$$\begin{vmatrix} -\lambda & 1 \\ 1 & -\lambda \end{vmatrix} = 0 \qquad (6.86)$$

which generates the characteristic equation:

$$\lambda^2 - 1 = 0 \qquad (6.87)$$

with eigenvalues $\lambda_1 = 1$ and $\lambda_2 = -1$. Inserting $\lambda_1 = 1$ into the first equation in 6.86 we have:

$$\begin{pmatrix} 0 & 1 \\ 1 & 0 \end{pmatrix} \begin{pmatrix} u_1 \\ u_2 \end{pmatrix} = 1 \cdot \begin{pmatrix} u_1 \\ u_2 \end{pmatrix} \qquad (6.88)$$

which gives the condition $u_1 = u_2$. The eigenvector $|u\rangle$ corresponding to the first eigenvalue $\lambda_1 = 1$ is therefore:

$$|u\rangle = \alpha \begin{pmatrix} 1 \\ 1 \end{pmatrix} \qquad (6.89)$$

where alpha is a real number. If we want to normalize to 1 this eigenvector we need to impose that $\langle u|u\rangle = 1$, i.e. $\alpha^2(1 + 1) = 1 \Rightarrow \alpha = \frac{1}{\sqrt{2}}$ and the first normalized eigenvector is:

$$|u\rangle = \frac{1}{\sqrt{2}} \begin{pmatrix} 1 \\ 1 \end{pmatrix}. \qquad (6.90)$$

---

[11] Also called *secular equation*.

The second eigenvector $|v\rangle$ is calculated in a similar way by inserting $\lambda_2 = -1$ into the first equation in 6.86. The second normalized eigenvector is:

$$|v\rangle = \frac{1}{\sqrt{2}} \begin{pmatrix} 1 \\ -1 \end{pmatrix}. \tag{6.91}$$

Let's discuss the case of **degenerate eigenvalues**, i.e. those eigenvalues who have associated more than one linearly independent eigenvector. As usual, we restrict ourselves to Hermitian matrices since they are the most important in quantum mechanics.

Given an $N \times N$ Hermitian matrix A spanning an N-dimensional linear vector space, let's suppose that the two eigenvectors $|a_1\rangle$ and $|a_2\rangle$ are corresponding to the same eigenvalue λ. This means that we have two independent relations:

$$\begin{aligned} A|a_1\rangle &= \lambda |a_1\rangle \\ A|a_2\rangle &= \lambda |a_2\rangle. \end{aligned} \tag{6.92}$$

Because $c_1|a_1\rangle + c_2|a_2\rangle$ is also an eigenvector of A, then the two eigenvectors associated with λ span a 2-dimensional sub-space of the N-dimensional space of all the eigenvectors. More in general, if we have M independent eigenvectors associated with the same eigenvalue, the eigenvectors span an N-dimensional sub-space, where $N < M$.

In general, the M eigenvectors associated with the degenerate eigenvalue are not orthonormal. In this case we can always build an orthonormal set starting with the degenerate eigenvector by using the so called *Gram-Schmidt* orthogonalization procedure. This procedure converts the degenerate eigenvectors $|a_i\rangle$, where $i = 1, 2, ..., M$ into an orthonormal set $|u_i\rangle$, where is also $i = 1, 2, ..., M$. Let's start with the set $|a_i\rangle$ of the degenerate eigenvector and we want to find a procedure to change this set into an orthonormal set $|u_i\rangle$, i.e. a set such that:

$$\langle u_i | u_j \rangle = \delta_{ij}, \quad i, j = 1, 2, ..., M. \tag{6.93}$$

We first generate an orthogonal set $|n_i\rangle$ and then into the final orthonormal set $|u_i\rangle$. The algorithm for the orthogonal set goes like:

$$\begin{aligned} |n_1\rangle &= |a_1\rangle \\ |n_2\rangle &= |a_2\rangle - \frac{\langle n_1 | a_2 \rangle}{\langle n_1 | n_1 \rangle} |n_1\rangle \\ |n_3\rangle &= |a_3\rangle - \frac{\langle n_1 | a_3 \rangle}{\langle n_1 | n_1 \rangle} |n_1\rangle - \frac{\langle n_2 | a_3 \rangle}{\langle n_2 | n_2 \rangle} |n_2\rangle. \end{aligned} \tag{6.94}$$

and so on. In general, we have:

$$|n_i\rangle = \left[I - \sum_{j=1}^{i-1} \frac{|n_j\rangle \langle n_j|}{\langle n_j | n_j \rangle} \right] |a_i\rangle, \quad i = 1, 2, ..., M. \tag{6.95}$$

A standard normalization will transform the orthogonal basis $|n\rangle$ into an orthonormal basis $|o\rangle$ by simply dividing each vector by its norm $\sqrt{\langle n|n\rangle}$.

We now have all the tools to diagonalize a Hermitian matrix B[12].

We need to state the conditions for a matrix A to be diagonalizable. First, the matrix needs to be square, i.e. same number of columns and rows. Second, the matrix needs to be similar to a diagonal matrix. Eq. 6.77 give the definition of *normal* for a Hermitian matrix. In general, a matrix A is similar to a diagonal matrix D if it exists an invertible matrix S such that:

$$D = S^{-1}AS. \tag{6.96}$$

Third, an $N \times N$ matrix is diagonalizable if and only if the dimension of its eigenspace, i.e. the number of linearly independent eigenvectors[13], is equal to N.

Once we have verified that A can be diagonalized, we need a procedure to build the matrix S and its inverse S^{-1}. Eq. 6.96 can be written as:

$$S^{-1}AS = \begin{pmatrix} \lambda_1 & 0 & \cdots & 0 \\ 0 & \lambda_2 & \cdots & 0 \\ 0 & 0 & \ddots & 0 \\ 0 & 0 & \cdots & \lambda_N \end{pmatrix}. \tag{6.97}$$

Let's multiply eq. 6.97 from the left by S. We have:

$$AS = S \begin{pmatrix} \lambda_1 & 0 & \cdots & 0 \\ 0 & \lambda_2 & \cdots & 0 \\ 0 & 0 & \ddots & 0 \\ 0 & 0 & \cdots & \lambda_N \end{pmatrix}. \tag{6.98}$$

Let's write the matrix S as a *block matrix*, i.e. a matrix where the columns are vectors:

$$S = \begin{pmatrix} \vdots & \vdots & \cdots & \vdots \\ |u_1\rangle & |u_2\rangle & \cdots & |u_N\rangle \\ \vdots & \vdots & \cdots & \vdots \end{pmatrix} = (|u_1\rangle \, |u_2\rangle \, \cdots \, |u_N\rangle). \tag{6.99}$$

where we have defined the kets as column vectors:

$$|u_1\rangle = \begin{pmatrix} u_1^1 \\ u_1^2 \\ \vdots \\ u_1^N \end{pmatrix}, \ |u_2\rangle = \begin{pmatrix} u_2^1 \\ u_2^2 \\ \vdots \\ u_2^N \end{pmatrix}, \ \cdots \ |u_N\rangle = \begin{pmatrix} u_N^1 \\ u_N^2 \\ \vdots \\ u_N^N \end{pmatrix}. \tag{6.100}$$

---

[12]Here we consider Hermitian matrices mainly because of their predominant role in quantum mechanics. More in general, a square matrix A is diagonalizable if it is similar to a diagonal matrix, i.e. if it is possible to construct an invertible matrix P such that $P^{-1}AP$ is a diagonal matrix.

[13]If some eigenvalues are degenerate we still need to count the number of linearly independent eigenvectors corresponding to the same eigenvalue.

We can therefore rewrite eq. 6.98 as:

$$A\,|u_i\rangle = \lambda_i\,|u_i\rangle,\quad (i = 1, 2, \cdots, N) \tag{6.101}$$

which gives the recipe to build the matrix S, i.e. the matrix built by placing in order the eigenvectors of the matrix A corresponding to their eigenvalues. The linear independence of the eigenvectors suggests that the matrix S is invertible. If the matrix A is Hermitian, then its eigenvectors can be chosen to be orthonormal and the matrix S is a unitary matrix, usually indicated with U for which $U^\dagger = U^{-1}$.

Let's show an example of diagonalization of a 2×2 matrix $A = \begin{pmatrix} 1 & 1 \\ 0 & 0 \end{pmatrix}$. The characteristic equation is obtained by setting:

$$\begin{vmatrix} 1 - \lambda & 1 \\ 0 & -\lambda \end{vmatrix} = 0 \tag{6.102}$$

with the corresponding secular equation:

$$-\lambda(1 - \lambda) = 0 \tag{6.103}$$

which gives the two eigenvalues $\lambda_1 = 0$ and $\lambda_2 = 1$. The eigenvectors are found by solving the eigenvalue equations separately for the two eigenvalues. For $\lambda_1 = 0$ we have:

$$A\,|u\rangle = \lambda_1\,|u\rangle$$
$$\begin{pmatrix} 1 & 1 \\ 0 & 0 \end{pmatrix}\begin{pmatrix} u_1 \\ u_2 \end{pmatrix} = 0 \tag{6.104}$$
$$u_1 + u_2 = 0.$$

Therefore, the first eigenvector $|u\rangle$ is:

$$|u\rangle = \alpha \begin{pmatrix} 1 \\ -1 \end{pmatrix} \tag{6.105}$$

where α is a multiplicative constant that we can set equal to 1. Remember that only the "direction" of wavefunctions is important.

For $\lambda_2 = 1$ we have:

$$A\,|v\rangle = \lambda_2\,|v\rangle$$
$$\begin{pmatrix} 1 & 1 \\ 0 & 0 \end{pmatrix}\begin{pmatrix} v_1 \\ v_2 \end{pmatrix} = \begin{pmatrix} v_1 \\ v_2 \end{pmatrix} \tag{6.106}$$
$$v_1 + v_2 = v_1.$$

The second eigenvector $|v\rangle$ is:

$$|v\rangle = \beta \begin{pmatrix} 1 \\ 0 \end{pmatrix} \tag{6.107}$$

where β is a multiplicative constant that we can set equal to 1 as for the other eigen-value.

Having found the eigenvector we can now build the matrix S according to the recipe 6.99:

$$S = \begin{pmatrix} 1 & 1 \\ -1 & 0 \end{pmatrix}. \tag{6.108}$$

The inverse of S is calculated according to 6.56 and it gives:

$$S^{-1} = \frac{1}{\det\{S\}} \begin{pmatrix} 0 & -1 \\ 1 & 1 \end{pmatrix} = \begin{pmatrix} 0 & -1 \\ 1 & 1 \end{pmatrix}. \tag{6.109}$$

The diagonalization is achieved by the following matrix multiplication:

$$D = S^{-1}AS = \begin{pmatrix} 0 & -1 \\ 1 & 1 \end{pmatrix} \begin{pmatrix} 1 & 1 \\ 0 & 0 \end{pmatrix} \begin{pmatrix} 1 & 1 \\ -1 & 0 \end{pmatrix} = \begin{pmatrix} 0 & 0 \\ 0 & 1 \end{pmatrix}. \tag{6.110}$$

Notice that the matrix D has the two eigenvalues $\lambda_1 = 0$ and $\lambda_2 = 1$ on the diagonal and zero everywhere else.

Now suppose we have a matrix B operating on a certain vector $|u\rangle$ to obtain a different vector $|v\rangle$. Suppose that we operate another matrix A on the vector $|v\rangle$. The operation we just did is written as $A|v\rangle = A(B|u\rangle) = AB|u\rangle$ where the last identity means that first the matrix B operates on the vector $|u\rangle$ and then the matrix A operates on the resulting vector. The product of two operators is normally another operator and we write $C|u\rangle = AB|u\rangle$. In general, the operator obtained by reversing the order of the matrices $D|u\rangle = BA|u\rangle$ is different from C. The difference between the two operators is called **commutator** and is defined as:

$$[A, B] = AB - BA. \tag{6.111}$$

We have already introduced the commutator between two operators in eq. 5.95 and 5.96. We have seen that, if two operators do not commute, i.e. their commutator is not equal to zero, then the two physical quantities that they represent cannot be measured with infinite precision but are subject to Heisenberg Uncertainty Principle. The same is valid in the matrix formulation of quantum mechanics that we are discussing here.

On the other hand, if two operators commute then the two physical quantities that they represent can be measured simultaneously and are not subject to Heisenberg Uncertainty Principle. Let's prove this last statement first: given two Hermitian matrices A and B with a common orthonormal basis of eigenfunctions, then the commutator $[A, B] = 0$.

If A and B have a common orthonormal basis of eigenfunctions then we have that:

$$\begin{aligned} A|u\rangle &= \lambda|u\rangle \\ B|u\rangle &= \mu|u\rangle \end{aligned} \tag{6.112}$$

where $|u\rangle$ is the common basis of eigenvectors. We have that:

$$AB|u\rangle = A\mu|u\rangle = \lambda\mu|u\rangle$$
$$BA|u\rangle = B\lambda|u\rangle = \mu\lambda|u\rangle \qquad (6.113)$$
$$(AB - BA)|u\rangle = 0.$$

The last equation in 6.113 is valid for a specific eigenvector $|u\rangle$. However, according to the spectral theorem 6.80, we can express any vector $|\psi\rangle$ as a linear combination of the eigenvectors. Therefore, we can write that $(AB - BA)|\psi\rangle = [A,B]|\psi\rangle = 0$, which is commonly written without the ket as $[A,B] = 0$.

The reverse can also be proved, i.e. given two matrices A and B, if their commutator is zero $[A,B] = 0$ then the two matrices share a common basis of eigenvectors. Let's first assume that there are no degenerate eigenvalues. We can write:

$$AB|u\rangle = BA|u\rangle = B\lambda|u\rangle = \lambda B|u\rangle. \qquad (6.114)$$

Let's write the first and last term of eq. 6.114:

$$A(B|u\rangle) = \lambda(B|u\rangle) \qquad (6.115)$$

which means that $B|u\rangle$ is itself an eigenfunction of A. If the eigenfunctions are not corresponding to degenerate eigenvalues then each eigenfunction corresponds to one distinct eigenvalue and we can write:

$$B|u\rangle = \mu|u\rangle \qquad (6.116)$$

thus proving that A and B have a common set of eigenvectors $|u\rangle$. If there is degeneracy in either matrices we know that using Gram-Schmidt orthogonalization procedure we can build an orthonormal set of eigenfunctions that will be shared by the two matrices with a line of reasoning similar to the case of distinct eigenvalues.

6.2.3 POSTULATES, PRINCIPLES AND THE PHYSICS OF MATRIX ALGEBRA

The mathematical machinery we have introduced in previous sections is connected to the physical world through a series of postulates first introduced by Dirac [12] and von Neumann in the years from 1930 to 1932. There are many equivalent versions of postulates of quantum mechanics and different books list postulates differently. We give here a typical list of postulates but we invite the reader to check other sources like, for example, the excellent book from Griffiths [19].

The postulates come in classes: the description of states, the description of physical quantities, the description of the act of measurement, the effect of the measurements on the state and the time evolution of a quantum system.

We assume that our physical system is nonrelativistic and is isolated. Postulate I goes as follows:

Postulate I: the state of a system at a certain time t is represented by a complex state vector $|\psi\rangle$, or equivalently by a complex wavefunction ψ, belonging to an

Hilbert space $\mathcal{H}$. The state vector or the wavefunction contains all the information about the quantum system in a probabilistic way. For example, if the wavefunction of a particle is a function of the continuous space coordinate x and of the time t, i.e. $\psi = \psi(x,t)$ according to Born, the quantity:

$$|\psi(x,t)|^2 dx \tag{6.117}$$

represents the probability of finding the particle between x and $x + dx$. In order to properly represent a probability, the wavefunction must be normalized:

$$\int_{-\infty}^{+\infty} |\psi(x,t)|^2 dx = 1 \tag{6.118}$$

stating the mere fact that the particle must be somewhere.

In Dirac notation, quantum states are identified with a ket $|\psi\rangle$ belonging to the Hilbert space $\mathcal{H}$. In general, two vectors $|\psi\rangle$ and $\alpha|\psi\rangle$, where α is a complex number, represent the same state. However, in quantum mechanics it is preferred to work with normalized vectors (or wavefunctions) so that $\langle\psi|\psi\rangle = 1$. The normalization of the vector still does not uniquely determine a state since we can have $|\alpha| = 1$. This means that two vectors differing by a complex phase factor of unit norm $e^{i\theta}$, i.e. $e^{i\theta}|\psi\rangle$ and $|\psi\rangle$, still represent the same quantum state.

Superposition Principle: a linear combination of states is a state. This is often referred as **quantum superposition** and is one of the fundamental principles[14] of quantum mechanics. The origin of this principle can be traced back to the linearity of Schrödinger equation. If $|\phi_1\rangle, |\phi_2\rangle, \cdots, |\phi_N\rangle$ are solutions of the Schrödinger equation, then $c_1|\phi_1\rangle + c_2|\phi_2\rangle + \cdots + c_N|\phi_N\rangle$ is also a valid solution.

Postulate II: Given a system, to each observable property is associated a Hermitian operator. Example of observables are: position, momentum, angular momentum, energy, etc. If the commutator between two observables A and B is zero, $[A,B] = 0$, then the two observables can be measured simultaneously with infinite precision and the observables are said to *commute*. If the two observables have a nonzero commutator, $[A,B] \neq 0$, then the two observables cannot be measured with infinite precision but are subject to the Heisenberg Uncertainty Principle. For example, position $\hat{a}$ and momentum $\hat{p}$ satisfy $[\hat{x}, \hat{p}_x] = i\hbar$.

Postulate III: Given a Hermitian operator A representing an observable, the act of measurements at the time t_0 will result in one of the eigenvalues a_i of A. At times $t > t_0$ the wavefunction describing the system **collapses** into the eigenvector $|a_i\rangle$ corresponding to the eigenvalue measured. Before the act of measurements, the system is in a state of superposition among all the possible states and it is not possible to predict *with certainty* the outcome of a measurements. After the act of measurements

---

[14]We list here the superposition principle as a postulate which is not correct. There is a lot of confusion in the physics literature among the terms *axiom*, *postulate* and *principle*. In the context of this book we consider axioms and postulates as fundamental assumptions that cannot be derived and they are the basis on which the theoretical framework of quantum mechanics is constructed. Principles are fundamental statements that can be derived.

we now have knowledge of the system, for example a specific particle has a certain momentum p_0, and any subsequent measurements of the momentum must return the same value p_0, in order to ensure reproducibility of the measurements process. If the system is in a state that cannot be expressed as a superposition of other states we say that the system is in a **pure state**. Pure states are also called wavefunctions.

Postulate IV: The probability P of obtaining a certain eigenvalue a_i after a measurements on a system in a state $|\psi_0\rangle$ is:

$$P = |\langle a_i|\psi_0\rangle|^2 \tag{6.119}$$

where $|\psi_0\rangle$ is the state vector describing the system at the time $t = t_0$.

It is important to make a distinction between a) measuring an observable that is originally in a state of superposition and as a result of the measurement act collapses into an eigenstate; and b) measuring an observable that is in a superposition state and repeat the measurements if the system is still in a superposition state. The first case a) is covered in Postulate III and we have already seen that every subsequent measurement will give the same eigenvalue a_i corresponding to the eigenvector resulting from the collapse of the state into an eigenstate. In other words, if a system is described by a state vector that is an eigenvector, then there is *certainty* about the result of the measurements. The first time a measurement is taken we can only calculate the probability of obtaining a certain result according to 6.119. However, every successive measurement will have probability 1 to obtain the same result.

The second case b) concerns a completely different experimental condition like, for example, a collection of N identical systems all described by the same state vector which is not an eigenvector. In this case we can only make statistical prediction and calculate the expectation value of the observables. As we already have shown in eq. 5.89 the expectation value of an operator A is:

$$\langle \hat{A} \rangle = \int \psi^* \hat{A} \psi dx \tag{6.120}$$

or, in Dirac notation:

$$\langle A \rangle = \langle \psi|A|\psi\rangle . \tag{6.121}$$

Postulate V: The state of a system evolves according to the Schrödinger equation $i\hbar\frac{\partial|\psi\rangle}{\partial t} = H|\psi\rangle$, where H is the Hamiltonian operator.

There is an equivalent enunciation of this postulate: the time evolution of a system is described by a unitary transformation $U(t,t_0)$ which give the state $|\psi(t)\rangle$ of the system at time t when the state of the system $|\psi(t_0)\rangle$ at time $t_0 < t$ is given:

$$|\psi(t)\rangle = U(t,t_0)|\psi(t_0)\rangle . \tag{6.122}$$

6.2.4 STERN-GERLACH EXPERIMENT

There are many physical situations that can be described by a simple two states system. The polarization states of a photon, either left and right circular polarization or vertical and horizontal polarization are a well-known example. The two-level description of a laser or the half-integer spin of a class of particles called *fermions* are other examples. We will concentrate on this last example by describing a classic experiment by Stern and Gerlach[15].

The Stern and Gerlach experiment [18] described in this section is an example of a good experiment built for the wrong reasons. We have seen in a previous chapter that Niels Bohr came up with a model of the atom where electrons orbit around the nucleus only in a set of permitted orbits. In order to make his model work with experimental data available to him, Bohr set up three postulates: electrons moves only on permitted orbits where they cannot emit e.m. radiation; electrons can jump from one orbit to the other via emission/absorption of photons of the right frequency; the angular momentum of the electron due to its orbit around the nucleus is an integer multiple of Planck constant $\hbar$. Bohr made the simplest assumption about the shape of electron's orbits: they are all circular. In 1916, Arnold Sommerfeld[16] made a refinement to Bohr's theory by allowing the orbits to be elliptical. This resulted in the introduction of 3 quantum numbers n, k and m corresponding, respectively, to Bohr's principal quantum number n, $k = 1, ..., n$ the shape of the orbit ($k = n$ for circular orbit) and $m = -k, -k + 1, ..., k$ the space orientation of the orbit. The space orientation corresponds to the projection of the vector angular momentum along a prescribed axis (for example the axis of an external magnetic field).

Stern and Gerlach wanted to verify the existence of quantized spatial orbits, i.e. those electron orbits that were allowed only a limited (quantized) range of spatial orientations. Stern and Gerlach modified the poles of an electromagnet in such a way that an inhomogeneous magnetic field is produced (see fig. 6.2). If a dipole is aligned along the gradient of a magnetic field as in fig. 6.2, the N pole of the magnetic dipole will feel a stronger force of attraction to the S pole of the electromagnet because of the gradient. Therefore, if a beam of atoms is forced to pass through the inhomogeneous magnetic field, and if the orbits are spatially quantized, then a beam splitting can occur as depicted in fig. 6.3. A magnetic dipole, immersed in a magnetic field has a potential energy of $V = -\mu \cdot \vec{B}$. If the magnetic field is not constant then there will be a force along the direction $N - S$ of the electromagnet equal to:

$$F_z = -\nabla(-\mu \cdot B) = \mu_z \frac{\partial B}{\partial z}. \qquad (6.123)$$

A magnetic dipole is produced in atoms because the orbiting electrons obey Ampere's law, i.e. the orbiting electron is seen as a current I in a loop of area A equal to the area of the orbit. Ampere's law generates a magnetic moment $\mu = IA$. The current generated by an electron orbiting can be estimated by calculating how much charge e, the charge of the electron, per unit second goes around the orbit. The electron move at a speed v and goes around the orbit $v/2\pi r$ times per second, where r is

---

[15]Otto Stern proposed the experiment in 1921. Walther Gerlach successfully built the equipment needed and made the measurements in 1922.

[16]Arnold Johannes Wilhelm Sommerfeld (1868-1951) was a German physicist who made important contribution to quantum and atomic physics.

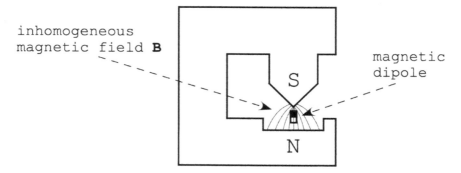

Figure 6.2 An inhomogenous magnetic field is generated between a pointed and a flat pole of an electromagnet. The magnetic lines of force are more concentrated close to the pointed pole thus generating a spatial gradient of the magnetic field B. A small magnetic dipole with its N pole up will experience a net force toward the S pointed pole because it is experiencing a slightly stronger force. If it is instead reversed with its S pole up, it will experience a net force downwards toward the N pole.

the radius of the orbit[17]. The current is therefore $I = ev/2\pi r$ The magnetic moment will be:

$$\mu = IA = \pi r^2 \cdot \frac{ev}{2\pi r} = \frac{1}{2}erv = \left(\frac{e}{2m_e}\right)L \qquad (6.124)$$

where $L = mrv$ is the angular momentum of the electron of charge e and mass m_e. Bohr and Sommerfeld theory requires that the angular momentum is quantized $L = 0, \hbar, 2\hbar, \ldots$ and when measured along the $z-$direction of the magnetic field it would be $L_z = m\hbar$. The smallest magnetic dipole is the so called *Bohr magneton* μ_B obtained by inserting $L = \hbar$ in eq: 6.124:

$$\mu_B = \frac{e\hbar}{2m_e}. \qquad (6.125)$$

Sommerfeld modifications of Bohr's theory was seen favorably because it gave a solutions to a few anomalous effects like the Zeeman effect and the Stark effects[18]. The new theory made the prediction that the number of lines visible after the splitting due to a magnetic field was equal to $2m + 1$ where m, defined above, is the integer value for the orbital angular momentum. This formula suggests that the spectral lines *always* split into an odd number of lines. The anomalous Zeeman effect instead produced both odd and even number of lines. This was the first hint that the angular momentum can assume half-integer values.

---

[17]Notice that this line of reasoning is purely classical because Heisenberg uncertainty principle does not allow the simultaneous knowledge of the radius of the electron's orbit and its velocity.

[18]The Zeeman effect is the splitting of a spectral lines into two or more lines in the presence of an external static magnetic field. The Stark effect is the splitting of spectral lines analogous to Zeeman effect but due an external electric field rather than a magnetic field.

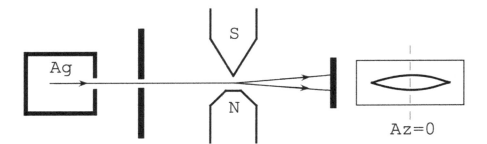

Figure 6.3 An atomic beam of silver atoms is collimated on an inhomogeneous magnetic field generated by an electromagnet. A splitting of the beam into two components is observed due to the intrinsic magnetic moment, or spin, of the external valence electron.

Stern and Gerlach set up the experiment to test Sommerfeld prediction of quantization of angular momentum. The experiment is briefly sketched in fig. 6.3: an oven on the left heats up silver metal until it melts and vaporizes. The atoms thermally acquire some energy and leave the oven through a small aperture and then they are further collimated by a collimator. The thin beam of silver atoms travels in high vacuum and is then subject to the inhomogeneous magnetic field where the force 6.123 will split the beam according to the orientation of the magnetic moment of each atom. The resulting beams are then deposited on a photographic emulsion where the deposition of silver atoms is observed.

Stern and Gerlach used silver atoms because they could be readily detected using a photographic emulsion and also because they have a single valence electron on the most external orbit. In this orbit the electron moves in the Coulomb potential caused by the 47 protons of the nucleus shielded by the 46 inner electrons. They believed that all the electrons on the inner orbits had a total magnetic moment equal to zero.

Based on the knowledge at the time of the experiment, if the interaction between the atomic magnetic dipoles and the external magnetic field was governed by classical physics, then they would have observed a fuzzy blob corresponding to all possible orientations of the dipoles but no splitting. Classically, the dipoles would be subject to Larmor precession in addition to the force due to the inhomogeneous external magnetic field. Being originated thermally, we would expect that all possible orientations are present in the silver beam.

On the other hand, if Bohr and Sommerfeld were right, then they would observe the beam splitting into three blobs corresponding to the $2m + 1$ states of the projection of the angular momentum along the z–axis. For angular momentum equal zero (L=0), they would observe one spot ($m = 0$) while for angular momentum equal to 1 ($m = -1, 0, +1$) they would observe three spots. In reality, instead of spots they would observe lines corresponding to the continuous variation of the azimuthal angle spreading.

The right side of fig. 6.3 shows what Stern and Gerlach observed: instead of one or three spots along the $Az = 0$ line, they observed a splitting into two corresponding to a half integer value for m. They also gave a value for the magnetic moment equal to $\pm\hbar/2$.

The proper explanation was given in 1925 by Samuel Goudsmit and George Uhlenbeck [42] which postulated that the outer orbit of the silver atom had $L = 0$ and that the electron possesses an **intrinsic** magnetic moment what was later called **spin**. When measured, the atoms always split into two beams: one corresponding to an intrinsic magnetic moment equal to $+\hbar/2$ deflecting the beam along the gradient of the static magnetic field and the other corresponding to an intrinsic magnetic moment equal to $-\hbar/2$ deflecting the beam in the opposite direction with respect to the gradient of the static magnetic field. Therefore, Stern and Gerlach performed the experiment to look for the wrong effect and instead they discovered the spin of the electron. Notice that they were not aware of the spin at the time of designing and building their experiment.

The convincing confirmation came later[34] in 1927 when hydrogen atoms were used which guaranteed that the atoms were effectively in their ground state $L = 0$.

In summary, the Stern and Gerlach experiment provided the first direct experimental evidence that angular momentum is quantized and revealed that the electron possesses an intrinsic magnetic moment which aligns always along the direction of the gradient of a static magnetic field. This is clearly against a classical interpretation of the spin: the silver atoms are thermally generated and it would be reasonable to assume that all the spins are randomly oriented. Instead, it seems that the spin is either "up" or "down" every time is measured no matter what direction the magnetic field is oriented.

6.2.5 ANGULAR MOMENTUM

In this section we show how quantum mechanics can properly describe the Stern and Gerlach experiment and justify the introduction of the three quantum numbers n, k and m corresponding, respectively, to Bohr's principal quantum number n, the shape of the orbit $k = 0, 1, ..., n$ and the space orientation of the orbit $m = -k, -k+1, ..., k$.

Let's start by briefly reviewing the classical angular momentum by discussing the motion of a particle described by the position vector r in Cartesian coordinate $r = (x, y, z)$, and the conjugate momentum $p = (p_x, p_y, p_z)$. If the particle is orbiting around the origin $r_0 = (0, 0, 0)$, classically there will be an orbital angular momentum given by the vector relation:

$$J = r \times p \qquad (6.126)$$

which is defining the angular momentum vector $J = (J_x, J_y, J_z)$. The three Cartesian components can therefore be written as:

$$
\begin{aligned}
J_x &= yp_z - zp_y \\
J_y &= zp_x - xp_z \\
J_z &= xp_y - yp_x.
\end{aligned}
\qquad (6.127)
$$

It is reasonable to make the transition to quantum mechanics by using eqs. 6.126 and 6.127 substituting the quantum mechanical operators for the position and the momentum. J_x, J_y and J_z are Hermitian operators whose eigenvalues are the measurable values of the angular momentum along the three directions x, y and z.

The corresponding quantum mechanical operator of the classical angular momentum is:

$$\hat{J} = -i\hbar(r \times \nabla) \tag{6.128}$$

with Cartesian coordinates:

$$\hat{J}_x = -i\hbar(y\frac{\partial}{\partial z} - z\frac{\partial}{\partial y})$$
$$\hat{J}_y = -i\hbar(z\frac{\partial}{\partial x} - x\frac{\partial}{\partial z}) \tag{6.129}$$
$$\hat{J}_z = -i\hbar(x\frac{\partial}{\partial y} - y\frac{\partial}{\partial x}).$$

Let us study the commutation relations of the operators J_i, where $i = x, y, z$. We have that, for x and y:

$$\begin{aligned}
[\hat{J}_x, \hat{J}_y] &= [(yp_z - zp_y), (zp_x - xp_z)] \\
&= (yp_z - zp_y)(zp_x - xp_z) - (zp_x - xp_z)(yp_z - zp_y) \\
&= yp_z(zp_x) - xp_z(zp_y) \\
&= -i\hbar \left[y\frac{\partial}{\partial z}(zp_x) - x\frac{\partial}{\partial z}(zp_y) \right] \\
&= i\hbar(xp_y - yp_x) \\
&= i\hbar\hat{J}_z.
\end{aligned} \tag{6.130}$$

The other commutators are obtained with a similar calculation:

$$\begin{aligned}
[\hat{J}_x, \hat{J}_y] &= i\hbar\hat{J}_z \\
[\hat{J}_y, \hat{J}_z] &= i\hbar\hat{J}_x \\
[\hat{J}_z, \hat{J}_x] &= i\hbar\hat{J}_y
\end{aligned} \tag{6.131}$$

which can be written in a compact form if we use $(1,2,3)$ instead of (x, y, z):

$$[\hat{J}_i, \hat{J}_j] = i\hbar\varepsilon_{ijk}\hat{L}_k \quad i = 1, 2, 3 \tag{6.132}$$

where ε_{ijk} is the Levi-Civita symbol, often referred as the completely antisymmetric tensor, defined in eq. 6.42 and we adopted Einstein's convention of summing over repeated indeces (k in this case). The Levi-Civita tensor is also used in the definition of vector product. For example, if $L = r \times p$, we can write for the components:

$$J_i = r_j p_k \varepsilon_{ijk}. \tag{6.133}$$

where repeated indices are assumed to be summed according to Einstein's convention. Using eq. 6.133, we can write the three equations in 6.131 as:

$$J \times J = i\hbar J. \tag{6.134}$$

The commutation relations 6.131 are at the basis of the quantum theory of angular momentum.

We have already encountered three operators with the same commutation relations when we introduced Pauli's matrices in eq. 6.18 hinting that they are connected to the (intrinsic) angular momentum of the electron.

The ket $|J\rangle$ defining the angular momentum is written using eq. 6.129:

$$|J\rangle = \begin{pmatrix} J_x \\ J_y \\ J_z \end{pmatrix}. \tag{6.135}$$

An explicit matrix form for the angular momentum operators 6.129 could be:

$$J_x = \frac{1}{\sqrt{2}} \begin{pmatrix} 0 & \hbar & 0 \\ \hbar & 0 & \hbar \\ 0 & \hbar & 0 \end{pmatrix}$$

$$J_y = \frac{1}{\sqrt{2}} \begin{pmatrix} 0 & -i\hbar & 0 \\ \hbar & 0 & -i\hbar \\ 0 & \hbar & 0 \end{pmatrix} \tag{6.136}$$

$$J_z = \frac{1}{\sqrt{2}} \begin{pmatrix} \hbar & 0 & 0 \\ 0 & 0 & 0 \\ 0 & 0 & -\hbar \end{pmatrix}$$

which satisfy the commutation relation 6.131.

The ket $|J\rangle$ defined in 6.135 has as components the three matrices defined in 6.136 and therefore is defined in a rather complicated Hilbert space. If we want to work with an angular momentum operator that is itself a 3×3 matrix defined in the same Hilbert space as the J operators in 6.136 we can use:

$$J^2 = \langle J|J\rangle = J_x^2 + J_y^2 + J_z^2 = 2\hbar^2 \begin{pmatrix} 1 & 0 & 0 \\ 0 & 1 & 0 \\ 0 & 0 & 1 \end{pmatrix} \tag{6.137}$$

which can be shown to be true after a lot of matrix multiplications. It can be shown that $[J^2, J_z] = 0$. In fact:

$$\begin{aligned} [J^2, J_z] &= [J_x^2 + J_y^2 + J_z^2, J_z] \\ &= [J_x^2, J_z] + [J_y^2, J_z] + [J_z^2, J_z] \\ &= [J_x^2, J_z] + [J_y^2, J_z] \end{aligned} \tag{6.138}$$

where we used one of the commutator properties $[A,A] = 0$ (see 5.97). Using another commutator property $[AB,C] = A[B,C] + [A,C]B$, let's calculate the two terms in 6.138:

$$[J_x^2, J_z] = [J_x \cdot J_x, J_z]$$
$$= J_x[J_x, J_z] + [J_x, J_z]J_x$$
$$= -i\hbar(J_xJ_y + J_yJ_x) \tag{6.139}$$

$$[J_y^2, J_z] = J_y[J_y, J_z] + [J_y, J_z]J_y$$
$$= i\hbar(J_yJ_x + J_xJ_y).$$

from which it follows that:

$$[J^2, J_z] = [J_x^2, J_z] + [J_y^2, J_z] = 0. \tag{6.140}$$

More in general, it can be shown that J^2 also commutes with J_x and J_y, i.e. $[J^2, J_{x,y,z}] = 0$.

Since J^2 commutes with J_z they have a common set of eigenbasis and therefore can be measured simultaneously with any precision.

In general, J^2 and J_z will produce different eigenvalues when expressed in the common eigenbasis. This means that the common eigenbasis need to be specified by two dimensionless parameters j and m and we can list the eigenbasis with the kets $|j,m\rangle$ assumed to be of unit norm, i.e. $\langle j,m|j,m\rangle = 1$.

The parameter m is defined by the eigenvalue equation:

$$J_z|j,m\rangle = m\hbar|j,m\rangle \tag{6.141}$$

where m is a dimensionless real number because J_z is Hermitian and $\hbar$ is needed to balance the physical dimensions of eq. 6.141. The most general eigenvalue equation for the operator J^2 is:

$$J^2|j,m\rangle = f(j,m)\hbar^2|j,m\rangle \tag{6.142}$$

where $f(j,m)$ is a real function of the two parameters (j,m) and $\hbar^2$ is needed to balance the physical dimensions of eq. 6.142.

From the definition $J^2 = J_x^2 + J_y^2 + J_z^2$ we have that $J^2 - J_z^2 = J_x^2 + J_y^2$. Since both operators J^2 and J_z are Hermitian then the expectation value of their difference is ≥ 0. In fact, given a Hermitian operator $A = A^\dagger$, we have that:

$$\langle \psi|A^2|\psi\rangle = \langle \psi|A^\dagger A|\psi\rangle = \langle \psi|\psi\rangle \geq 0. \tag{6.143}$$

We therefore have:

$$\langle j,m|(J_x^2 + J_y^2)|j,m\rangle = \langle j,m|(J^2 - J_z^2)|j,m\rangle$$
$$= \langle j,m|(f(j,m)\hbar^2 - m^2\hbar^2)|j,m\rangle$$
$$= (f(j,m)\hbar^2 - m^2\hbar^2)\langle j,m|j,m\rangle \tag{6.144}$$
$$= f(j,m)\hbar^2 - m^2\hbar^2 \geq 0.$$

Therefore we must have:

$$m^2 \leq f(j,m). \tag{6.145}$$

Let's consider the operator $J^2 - J_z = J_x^2 + J_y^2$. In analogy to the reasoning that brought us to introduce the raising and lowering operator in the study of the quantum harmonic oscillator (see eq. 5.183), let's create two operators that factorize the sum $J_x^2 + J_y^2$:

$$J_{\pm} = J_x \pm iJ_y \tag{6.146}$$

where, as it was the case of the quantum harmonic oscillator, they are called ladder operators and, specifically raising operator J_+ and lowering operator J_-. They obey the following commutation relations:

$$[J^2, J_{\pm}] = 0$$
$$[J_z, J_{\pm}] = \pm \hbar J_{\pm}. \tag{6.147}$$

In fact, we have:

$$\begin{aligned}
J^2 J_{\pm} |j,m\rangle &= J^2 (J_x \pm iJ_y) |j,m\rangle \\
&= J^2 J_x |j,m\rangle \pm i J^2 J_y |j,m\rangle \\
&= J_x J^2 |j,m\rangle \pm J_y J^2 |j,m\rangle \\
&= \hbar^2 f(j,m) J_{\pm} |j,m\rangle
\end{aligned} \tag{6.148}$$

where the last equality holds because we have already shown that J^2 commutes with both J_x and J_y. Notice that the operators $J_{\pm}$ are not Hermitian and therefore do not represent an observable; instead, they are Hermitian conjugates, i.e. $J_+ = J_-^{\dagger}$ and $J_- = J_+^{\dagger}$.

Eq. 6.148 tells us that not only $J_{\pm} |j,m\rangle$ is an eigenstate of J^2, but that the eigenvalue of J^2 is not affected by the raising or lowering operators. It follows that the ladder operators do not change the magnitude of the angular momentum.

The same cannot be said for the operator J_z. In fact we have:

$$\begin{aligned}
J_z J_+ |j,m\rangle &= J_+ J_z |j,m\rangle + \hbar J_+ |j,m\rangle \\
&= (m+1)\hbar |j,m\rangle
\end{aligned} \tag{6.149}$$

where we have used the eigenvalue eq. 6.141 and the commutation relations 6.147. A similar calculation shows that:

$$J_z J_- |j,m\rangle = (m-1)\hbar |j,m\rangle \tag{6.150}$$

fully justifying the names of raising and lowering operators. The effect of $J_{\pm}$ acting on $|j,m\rangle$ is to "raise" (J_+) or "lower" (J_-) the eigenvalue of the eigenstate $|j,m\rangle$ when the operator J_z is applied. In other words, $J_{\pm} |j,m\rangle$ is an eigenvector of J_z with eigenvalue $(m \pm 1)\hbar$.

We can go a step further and state that, because of eqs. 6.149 and 6.150 we can write:

$$J_{\pm} |j,m\rangle = N_{\pm} |j,m \pm 1\rangle \tag{6.151}$$

where $N_\pm$ is a normalization constant calculated by evaluating the expectation value of $J_- J_+$:

$$\langle j,m|J_+ J_- |j,m\rangle = \langle j,m|\,(J^2 - J_z^2 - \hbar J_z)\,|j,m\rangle$$
$$= \hbar^2(f - m^2 - m)\,\langle j,m|j,m\rangle \qquad (6.152)$$

where we used eq. 6.155 below. With a similar calculation we can write:

$$\langle j,m|J_+ J_- |j,m\rangle = \hbar^2(f - m^2 + m)\,\langle j,m|j,m\rangle. \qquad (6.153)$$

Equating eqs. 6.152 and 6.153 to the norm of eq. 6.151 we obtain the normalization factors:

$$N_\pm = \hbar\,\sqrt{f - m^2 \mp m}. \qquad (6.154)$$

Let's go back to eq. 6.145. This equation tells us that, given a value of $f(j,m)$, in short f from now on, the value of m is bounded. This means that there exists a state $|j,m_{max}\rangle$ such that $J_+ |j,m_{max}\rangle = 0$. If we apply the operator J_- to the null state $J_+ |j,m_{max}\rangle$ we still have the null state.

Let's calculate the form of the operator $J_- J_+$:

$$J_- J_+ = (J_x - iJ_y)(J_x + iJ_y)$$
$$= J_x^2 + J_y^2 + iJ_x J_y - iJ_y J_x$$
$$= J_x^2 + J_y^2 + i[J_x, J_y] \qquad (6.155)$$
$$= J^2 - J_z^2 - \hbar J_z.$$

Let's apply the operator 6.155 to the eigenstate $|j,m_{max}\rangle$:

$$J_- J_+ |j,m_{max}\rangle = (J^2 - J_z^2 - \hbar J_z)|j,m_{max}\rangle$$
$$= (f\hbar^2 - m_{max}^2 \hbar^2 - \hbar^2 m_{max})|j,m_{max}\rangle \qquad (6.156)$$
$$= 0$$

which implies that:

$$f - m_{max}^2 - m_{max} = 0. \qquad (6.157)$$

We now repeat the same line of reasoning for the lowering operator J_-. It must exist a value m_{min} for which for which the application of the lowering operator produces the zero state. Similarly to what we have already done for the raising operator, let's first calculate the operator $J_+ J_-$:

$$J_+ J_- = (J_x + iJ_y)(J_x - iJ_y)$$
$$= J_x^2 + J_y^2 - i[J_x, J_y] \qquad (6.158)$$
$$= J^2 - J_z^2 + \hbar J_z$$

and let's apply it to the eigenstate $|j,m_{max}\rangle$. After a little bit of algebra we obtain:

$$f - m_{min}^2 + m_{min} = 0. \qquad (6.159)$$

Equating eqs. 6.157 and 6.159 we have:

$$m_{max}^2 + m_{max} - m_{min}^2 + m_{min} = 0. \tag{6.160}$$

Eq. 6.160 is a simple second-order algebraic equation that can be solved with respect to m_{max} to produce two solutions: $m_{max} = -m_{min}$ and $m_{max} = m_{min} - 1$. The second solution is not physical because we have shown already that m_{min} is the minimum value and m_{max} cannot be less than m_{min}.

The first solution $m_{max} = -m_{min}$ tells us that the spectrum of eigenvalues m is such that, to go from the minimum value m_{min} to the maximum value m_{max} we need to apply the raising operator n times, each time increasing the eigenvalue by $\hbar$. Notice that n is an integer index. We have:

$$2m_{max} = n \tag{6.161}$$

or equivalently:

$$m_{max} = \frac{n}{2}. \tag{6.162}$$

We can finally obtain the function $f(j, m)$ by inserting 6.162 into 6.157:

$$f = \left(\frac{n}{2}\right)\left(\frac{n}{2} + 1\right). \tag{6.163}$$

If we now make the substitution $j = \frac{n}{2}$ where $n = 0, 1, \ldots$ is an integer, we have that the two eigenvalue equations 6.141 and 6.142 become:

$$\begin{aligned} J^2|j, m\rangle &= j(j+1)\hbar^2|j, m\rangle \\ J_z|j, m\rangle &= m\hbar|j, m\rangle. \end{aligned} \tag{6.164}$$

and eq. 6.151 becomes:

$$J_\pm|j, m\rangle = \hbar\sqrt{j(j+1) - m(m \pm 1)}|j, m \pm 1\rangle. \tag{6.165}$$

In summary, because of the integer index $n = 0, 1, \ldots$; the index j assumes alternatively integer and semi-integer values $j = 0, \frac{1}{2}, 1, \frac{3}{2}, \ldots$; the index m assumes $2j + 1$ values and is bounded in the interval $(-j, j)$ according to $m = -j, -j + 1, \ldots, j - 1, j$.

In summary, we have seen that a quantum theory of the angular momentum requires the introduction of two quantum numbers j and m. The theory then predicts that the angular momentum is quantized and that the eigenvalues of J^2, corresponding to the value of the square of the angular momentum, and J_z, corresponding to the z-component of J, also are quantized. Although the eigenvalues of the angular momentum J^2 are quantized, they are not bounded while the eigenvalues of its z-component J_z are bounded to be between $-j$ and j.

If we attempt at building a classical picture of the angular momentum we can associate the integer eigenvalues of the quantum number j with the **orbital** angular momentum usually indicated by L. This is exactly what Bohr has done with his hydrogen atom model and relative quantization rules. As we discussed in chapter 4,

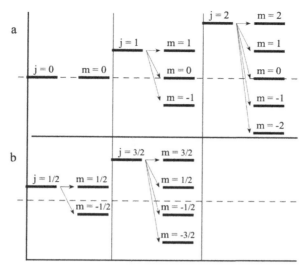

Figure 6.4 The splitting of integer values of the total angular momentum J for a) the three cases $j = 0, 1$ and 2, and b) two semi-integer cases $j = 1/2$ and $3/2$.

in order to explain the emission spectrum of hydrogen atoms, Bohr postulated that the angular momentum is quantized in integer units of $\hbar$ according to eq. 4.3. If we naively assume that the electron is a negative charge orbiting a positively charged proton, the orbital motion generates a magnetic dipole μ given by eq. 6.124. The smallest magnetic dipole generated by an electron is the Bohr magneton obtained by assuming that the smallest angular momentum is equal to $\hbar$.

As we have seen in section 6.2.4, Stern and Gerlach built an experiment aimed at observing the splitting of an atomic beam into three bands as predicted by assuming $j = 1$. Instead, Stern and Gerlach observed a splitting into two bands corresponding to $j = \frac{1}{2}$. In order to account for this case, it was assumed that the electron possesses a **total angular momentum** J equal to:

$$J = L + S \tag{6.166}$$

resulting from the sum of the orbital angular momentum L, due to the motion of the electron, and an intrinsic spin angular moment S. In complete analogy with the total angular momentum J, both L and S are Hermitian operators that obey commutation relations similar to eqs. 6.131 and 6.132:

$$[\hat{L}_i, \hat{L}_j] = i\hbar\varepsilon_{ijk}\hat{L}_k$$
$$[\hat{S}_i, \hat{S}_j] = i\hbar\varepsilon_{ijk}\hat{S}_k \tag{6.167}$$

where $i = 1, 2, 3$ and Einstein's convention is adopted intending that a summation on repeated indices is not explicitly written. Continuing with the analogy, if we build the operator L^2 and L_z, we see that they commute:

$$[L^2, L_z] = 0 \tag{6.168}$$

thus allowing a set of common eigenvectors $|\ell, m\rangle$ with the eigenvalue equations:

$$L^2 |\ell, m\rangle = \ell(\ell+1)\hbar^2 |\ell, m\rangle$$
$$L_z |\ell, m\rangle = m\hbar |\ell, m\rangle \tag{6.169}$$

where ℓ is an integer and m is bound to the $(2\ell+1)$ values $-\ell, -\ell+1, ..., \ell$.

Before studying the properties of the spin, we want to discuss more in detail the origin of the discrete nature of the eigenvalues of the angular momentum, in particular the L_z components. We first discuss the standard derivation of the discrete spectrum and then we give a different derivation based on less assumptions which is, in out opinion, more convincing.

Let's rewrite eq. 6.126 for the orbital angular momentum:

$$L = r \times p. \tag{6.170}$$

We begin the demonstration by expressing the operator $L_z = -i\hbar\frac{\partial}{\partial z}$ in spherical coordinates. The transformations from Cartesian to spherical coordinates are given by:

$$x = r\sin\theta\cos\phi$$
$$y = r\sin\theta\sin\phi \tag{6.171}$$
$$z = r\cos\theta$$

while the inverse transformations are given by[19]:

$$r = (x^2 + y^2 + z^2)^{\frac{1}{2}}$$
$$\theta = \tan^{-1}(\frac{\sqrt{x^2+y^2}}{z}) \tag{6.172}$$
$$\phi = \tan^{-1}\frac{y}{x}$$

where the geometry of the transformations is shown in fig. 6.5.

Let's show that the gradient operator, written in Cartesian coordinates as:

$$\nabla = \frac{\partial}{\partial x}e_x + \frac{\partial}{\partial y}e_y + \frac{\partial}{\partial z}e_z \tag{6.173}$$

in spherical coordinates, is written as:

$$\nabla = \frac{\partial}{\partial r}e_r + \frac{1}{r}\frac{\partial}{\partial\theta}e_\theta + \frac{1}{r\sin\theta}\frac{\partial}{\partial\phi}e_\phi \tag{6.174}$$

---

[19]This is the usual coordinate transformations written in physics books. Mathematicians tend to exchange θ with ϕ.

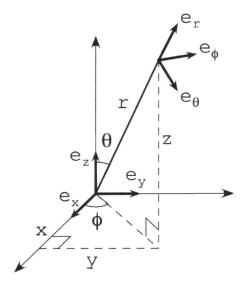

Figure 6.5 Geometry of the transformations between Cartesian and spherical coordinates.

where the three unit vectors e_r, e_θ, e_ϕ form an orthonormal basis for the spherical coordinates r, θ, ϕ defined in eq. 6.172 in analogy to the three unit vectors e_x, e_y, e_z defining the Cartesian coordinate system x, y, z (see fig. 6.5).

The next step consists in expressing the orthonormal basis e_r, e_θ, e_ϕ in terms of the orthonormal vectors e_x, e_y, e_z. An inspection of fig. 6.5 shows that e_r is in the same direction as the vector r. Therefore the unit vector e_r is expressed by:

$$e_r = \frac{r}{\|r\|} = \frac{xe_x + ye_y + ze_z}{\sqrt{x^2 + y^2 + z^2}} \tag{6.175}$$

where $\|r\| = \sqrt{x^2 + y^2 + z^2}$ represents the norm of the vector r. Inserting eq. 6.171 into eq. 6.175 we have:

$$e_r = \sin\theta\cos\phi e_x + \sin\theta\sin\phi e_y + \cos\theta e_z. \tag{6.176}$$

Since the angle ϕ is defined in the $xy-$ plane, the unit vector e_ϕ must be parallel to the $xy-$plane. This means that the unit vector e_ϕ has a null z component and must be of the form $ae_x + be_y$. If we let the angle θ assume the special value $\pi/2$, then the vector e_r is forced to lay on the $xy-$plane and assumes the form $e_r = \cos\phi e_x + \sin\phi e_y$. It immediately follows that the unit vector e_ϕ perpendicular to e_r oriented along the positive ψ direction is:

$$e_\phi = -\sin\phi e_x + \cos\phi e_y. \tag{6.177}$$

The third and final unit vector can be calculated using the condition $e_\theta = e_\phi \times e_r$ and we get:

$$e_\theta = \cos\theta\cos\phi e_x + \cos\theta\sin\phi e_y - \sin\theta e_z. \tag{6.178}$$

After some lengthy algebra, eqs. 6.176, 6.177 and 6.178 can be inverted to give:

$$e_x = \sin\theta\cos\phi\, e_r - \sin\phi\, e_\phi + \cos\theta\cos\phi\, e_\theta$$
$$e_y = \sin\theta\sin\phi\, e_r + \cos\phi\, e_\phi + \cos\theta\sin\phi\, e_\theta \qquad (6.179)$$
$$e_z = \cos\theta\, e_r - \sin\theta\, e_\theta.$$

Let's write down the chain rule for the relevant partial differentials:

$$\frac{\partial}{\partial r} = \frac{\partial}{\partial x}\frac{\partial x}{\partial r} + \frac{\partial}{\partial y}\frac{\partial y}{\partial r} + \frac{\partial}{\partial z}\frac{\partial z}{\partial r}$$
$$\frac{\partial}{\partial\theta} = \frac{\partial}{\partial x}\frac{\partial x}{\partial\theta} + \frac{\partial}{\partial y}\frac{\partial y}{\partial\theta} + \frac{\partial}{\partial z}\frac{\partial z}{\partial\theta} \qquad (6.180)$$
$$\frac{\partial}{\partial\phi} = \frac{\partial}{\partial x}\frac{\partial x}{\partial\phi} + \frac{\partial}{\partial y}\frac{\partial y}{\partial\phi} + \frac{\partial}{\partial z}\frac{\partial z}{\partial\phi}$$

which, using eqs. 6.171 yields:

$$\frac{\partial}{\partial r} = \sin\theta\cos\phi\frac{\partial}{\partial x} + \sin\theta\sin\phi\frac{\partial}{\partial y} + \cos\theta\frac{\partial}{\partial z}$$
$$\frac{\partial}{\partial\theta} = r\cos\theta\cos\phi\frac{\partial}{\partial x} + r\cos\theta\sin\phi\frac{\partial}{\partial y} - r\sin\theta\frac{\partial}{\partial z} \qquad (6.181)$$
$$\frac{\partial}{\partial\phi} = -r\sin\theta\sin\phi\frac{\partial}{\partial x} + r\sin\theta\cos\phi\frac{\partial}{\partial y}.$$

Eq. 6.181 can be inverted to give:

$$\frac{\partial}{\partial x} = \sin\theta\cos\phi\frac{\partial}{\partial r} + \frac{1}{r}\cos\phi\cos\theta\frac{\partial}{\partial\theta} - \frac{1}{r}\frac{\sin\phi}{\sin\theta}\frac{\partial}{\partial\phi}$$
$$\frac{\partial}{\partial y} = \sin\theta\sin\phi\frac{\partial}{\partial r} + \frac{1}{r}\sin\phi\cos\theta\frac{\partial}{\partial\theta} + \frac{1}{r}\frac{\cos\phi}{\sin\theta}\frac{\partial}{\partial\phi} \qquad (6.182)$$
$$\frac{\partial}{\partial z} = \cos\theta\frac{\partial}{\partial r} - \frac{1}{r}\sin\theta\frac{\partial}{\partial\theta}.$$

We have all the ingredients to write the gradient is spherical coordinates. In the expression of the gradient in Cartesian coordinates 6.173 we substitute the values from eqs. 6.182 and 6.179 to finally obtain:

$$\nabla = \frac{\partial}{\partial r}e_r + \frac{1}{r}\frac{\partial}{\partial\theta}e_\theta + \frac{1}{r\sin\theta}\frac{\partial}{\partial\phi}e_\phi. \qquad (6.183)$$

After some lengthy calculations[20] the Laplacian ∇^2 can be written as:

$$\nabla^2 = \frac{\partial^2}{\partial r^2} + \frac{2}{r}\frac{\partial}{\partial r} + \frac{\cos\theta}{r^2\sin\theta}\frac{\partial}{\partial\theta} + \frac{1}{r^2}\frac{\partial^2}{\partial\theta^2} + \frac{1}{r^2\sin^2\theta}\frac{\partial^2}{\partial\phi^2}. \qquad (6.184)$$

---

[20]The interested reader can find the details in any good textbook of mathematical methods for physics like, for example, Riley, Hobson and Bench [23].

We progress now in expressing the classical angular momentum 6.170 in terms of quantum operators. In terms of components, we have:

$$L = \begin{vmatrix} e_x & e_y & e_z \\ x & y & z \\ p_x & p_y & p_z \end{vmatrix} = (yp_z - zp_y)e_x + (zp_x - xp_z)e_y + (xp_y - yp_x)e_z \qquad (6.185)$$

which defines the three components of the angular momentum:

$$\begin{aligned} L_x &= (yp_z - zp_y) \\ L_y &= (zp_x - xp_z) \\ L_z &= (xp_y - yp_x). \end{aligned} \qquad (6.186)$$

The quantum mechanical operators, in Cartesian coordinates, are obtained by substituting $p = -i\hbar\nabla$ into 6.186:

$$\begin{aligned} L_x &= -i\hbar(y\frac{\partial}{\partial z} - z\frac{\partial}{\partial y}) \\ L_y &= -i\hbar(z\frac{\partial}{\partial x} - x\frac{\partial}{\partial z}) \\ L_z &= -i\hbar(x\frac{\partial}{\partial y} - y\frac{\partial}{\partial x}). \end{aligned} \qquad (6.187)$$

Using eqs. 6.182, and after some algebra, we can express the angular momentum components in spherical coordinates:

$$\begin{aligned} L_x &= i\hbar\left(\sin\phi\frac{\partial}{\partial\theta} + \cos\phi\cot\theta\frac{\partial}{\partial\phi}\right) \\ L_y &= i\hbar\left(-\cos\phi\frac{\partial}{\partial\theta} + \sin\phi\cot\theta\frac{\partial}{\partial\phi}\right) \\ L_z &= -i\hbar\frac{\partial}{\partial\phi}. \end{aligned} \qquad (6.188)$$

It is a matter of some algebra to obtain the following expressions in spherical coordinates for L:

$$L = -i\hbar \times \nabla = -i\hbar\left(e_\phi\frac{\partial}{\partial\theta} - e_\theta\frac{1}{\sin\theta}\frac{\partial}{\partial\phi}\right) \qquad (6.189)$$

and for L^2 and $L_\pm$:

$$\begin{aligned} L^2 &= L \cdot L = -\hbar^2\left[\frac{1}{\sin\theta}\frac{\partial}{\partial\theta}\left(\sin\theta\frac{\partial}{\partial\theta}\right) + \frac{1}{\sin^2\theta}\frac{\partial^2}{\partial\phi^2}\right] \\ L_\pm &= \pm\hbar e^{\pm i\phi}\left(\frac{\partial}{\partial\theta} \pm i\cot\theta\frac{\partial}{\partial\phi}\right) \end{aligned} \qquad (6.190)$$

Let's study the eigenvalue equation for the operator L_z when it is expressed in spherical coordinates. We have:

$$L_z \psi = \lambda \psi \tag{6.191}$$

where ψ are the eigenfunctions of L_z with eigenvalues λ. Using L_z from eq. 6.187 we have:

$$-i\hbar \frac{\partial \psi}{\partial \phi} = \lambda \psi \tag{6.192}$$

which admits as solution:

$$\psi = c e^{\frac{i\lambda\phi}{\hbar}} \tag{6.193}$$

where c is a normalization constant. Because ϕ is an angle defined between 0 and 2π, adding 2π to the angle ϕ should result in the same value of the eigenfunction ψ, i.e. $\psi(\phi + 2\pi) = \psi(\phi)$. We therefore must have:

$$c e^{\frac{i\lambda(\phi+2\pi)}{\hbar}} = c e^{\frac{i\lambda\phi}{\hbar}} \tag{6.194}$$

which simplifies into:

$$e^{\frac{i\lambda 2\pi}{\hbar}} = 1. \tag{6.195}$$

The exponential in eq. 6.195 is equal to 1 only if:

$$\lambda = m\hbar \tag{6.196}$$

where m must be an integer.

Since we already know that m is restricted by $m = -\ell, (-\ell+1), ..., \ell$ we deduce that ℓ must be an integer too. Therefore, orbital angular momentum eigenvalues are restricted to be integers and not half-integers like the total angular momentum J or the spin S.

The above derivation has an important assumption: the condition 6.194 assumes that the eigenfunctions ψ are *single-valued* under a rotation of the angle ϕ of 2π. This assumption is perfectly reasonable when dealing with observables like, for example, the fields in e.m. theory. In the case of quantum mechanics, the wavefunctions, and thus the eigenfunctions, are not observables: the observables are either of the form of a norm or an expectation value. In both cases the single-value condition is not as strong as in classical mechanics.

We conclude this section with an alternative proof[21] that the eigenvalues of the $z-$component of the orbital angular momentum are integer multiples of $\hbar$. A part from numerical coefficients, the orbital angular momentum operator can be written as a combination of Hermitian operators:

$$L_z = -i\hbar(Q_x P_y - Q_y P_x) \tag{6.197}$$

---

[21]This argument is taken from the book by Ballantine[4].

obeying the following commutation relations:

$$[Q_\alpha, Q_\beta] = i\hbar\delta_{\alpha,\beta}, \quad \alpha, \beta = x, y, z. \tag{6.198}$$

For the rest of this proof, we will use units where position and momentum operators are dimensionless together with redefining $\hbar = 1$. We can introduce the following operators:

$$
\begin{aligned}
q_1 &= \frac{1}{\sqrt{2}}(Q_x + P_y) \\
q_2 &= \frac{1}{\sqrt{2}}(Q_x - P_y) \\
p_1 &= \frac{1}{\sqrt{2}}(P_x + Q_y) \\
p_2 &= \frac{1}{\sqrt{2}}(P_x + Q_y).
\end{aligned}
\tag{6.199}
$$

These new operators obeys the same commutation relations obeyed by the components of the operator L:

$$[q_i, q_j] = i\delta_{i,j}, \quad [p_i, p_j] = i\delta{i,j}, \quad i, j = 1, 2. \tag{6.200}$$

Inserting 6.199 into 6.197 we obtain:

$$L_z = \frac{1}{2}(p_1^2 + q_1^2) - \frac{1}{2}(p_2^2 + q_2^2). \tag{6.201}$$

The above equation represents the difference between two uncoupled harmonic oscillators of unit mass and unit angular frequency. We have seen in section 5.10 that an harmonic oscillator has a discrete spectrum of energy eigenvalues according to eq. 5.200. Because of the commutation relations 6.200, the eigenvalues of the difference of two operators are the difference of the eigenvalues. We can therefore state that the eigenvalues of L_z can be written as:

$$\lambda = \left(n_1 + \frac{1}{2}\right)\hbar - \left(n_2 + \frac{1}{2}\right)\hbar = (n_1 - n_2)\hbar. \tag{6.202}$$

Eq. 6.202 tells us that the eigenvalues of L_z are the difference of two non negative integers times $\hbar$ thus providing another proof that the eigenvalues of the orbital angular momentum are integer multiples of $\hbar$ and finally justifying the original postulate from Bohr of eq. 4.3 which has been so successful in explaining the hydrogen atom.

6.2.6 SPHERICAL HARMONICS

In this section, we find an explicit form for the eigenstates of the commuting operators L^2 and L_z which are commonly referred to as *spherical harmonics*. Finding the eigenstates of the angular momentum is a long and tedious job and we give here a short version using the algebraic method that we have used already for the harmonic oscillator.

The eigenstates $|\ell, m\rangle$ of the angular momentum are generally called **Spherical Harmonics** and are indicated with the symbol $Y_{\ell,m}(\theta, \phi)$. By definition, eqs. 6.169 can be written as:

$$L^2 Y_{\ell,m}(\theta, \phi) = \ell(\ell+1)\hbar^2 Y_{\ell,m}(\theta, \phi)$$
$$L_z Y_{\ell,m}(\theta, \phi) = m\hbar Y_{\ell,m}(\theta, \phi). \tag{6.203}$$

Let's restrict ourselves for now to the special subset of spherical harmonics for which $\ell = m$ and apply the raising operator of eq. 6.190:

$$L_+ Y_{\ell,m}(\theta, \phi) = \hbar e^{i\phi} \left(\frac{\partial}{\partial \theta} + i\cot\theta \frac{\partial}{\partial \phi} \right) Y_{\ell,m}(\theta, \phi) = 0. \tag{6.204}$$

Eq. 6.204 is a consequence of the restrictions on the value of m not to exceed the value of ℓ being restricted to $m = -\ell, -\ell+1, ..., \ell$.

Eq. 6.204 can be solved by assuming that the eigenstate can be separated into the product of two functions each depending only on, respectively, the angular coordinate θ and ϕ:

$$Y_{\ell,m}(\theta, \phi) = \Theta_{\ell,\ell}(\theta)\Phi_\ell(\phi). \tag{6.205}$$

Inserting the expression 6.205 into 6.204 and after a bit of algebra we obtain:

$$\frac{1}{\cot\theta} \frac{1}{\Theta_{\ell,\ell}(\theta)} \frac{d\Theta_{\ell,\ell}}{d\theta} = \frac{1}{i\Phi_\ell(\phi)} \frac{d\Phi}{d\phi} = a = \text{constant} \tag{6.206}$$

where a is a constant. Eq. 6.206 then splits into two equations each depending only on one variable. The first equation is:

$$\frac{d\Phi_\ell}{\Phi_\ell} = ia \, d\phi \tag{6.207}$$

which admits the solution:

$$\Phi_\ell = A e^{ia\phi}. \tag{6.208}$$

The function Φ_ℓ is also an eigenstate of L_z for which:

$$L_z \Phi_\ell = m\hbar \Phi_\ell. \tag{6.209}$$

Using the third eq. 6.188, we have:

$$L_z \Phi_\ell = m\hbar \Phi_\ell = -i\hbar \frac{\partial \Phi_\ell ll}{\partial \phi} = a\hbar \Phi_\ell \tag{6.210}$$

from which it follows that $a = m$ and the eigenstate 6.208, for the special case $m = \ell$ is written as:

$$\Phi_\ell = A e^{i\ell\phi}. \tag{6.211}$$

We want normalize 6.211, i.e. its square modulus of to be equal to 1:

$$|\Phi_\ell|^2 = A^2 \int_0^{2\pi} \Phi_\ell^* \Phi_\ell d\phi = A^2 \int_0^{2\pi} e^{-i\ell\phi} e^{i\ell\phi} d\phi = A^2 \, 2\pi = 1 \qquad (6.212)$$

from which:

$$A = \frac{1}{\sqrt{2}}. \qquad (6.213)$$

From fig. 6.5 we see that the angle ϕ is defined between 0 and 2π meaning that a full rotation of 2π returns the eigenstate to itself, i.e. $\Phi_\ell(\phi) = \Phi_\ell(\phi + 2\pi)$. It follows that:

$$e^{i\ell 2\pi} = 1 \qquad (6.214)$$

implying that ℓ must be an integer. This is the case of *orbital angular momentum* and not the spin which will be treated in the next section.

For the θ coordinate, from eq. 6.206 we have:

$$\frac{1}{\cot\theta} \frac{1}{\Theta_{\ell,\ell}(\theta)} \frac{d\Theta_{\ell,\ell}}{d\theta} = \ell \qquad (6.215)$$

from which we have:

$$\frac{d\Theta_{\ell,\ell}}{\Theta_{\ell,\ell}} = \ell \cot\theta d\theta \qquad (6.216)$$

which integrated gives:

$$\Theta_{\ell,\ell} = C \sin^\ell \theta \qquad (6.217)$$

where C is a constant to be determined from the normalization of the eigenstate. After some algebra and combining the ϕ eigenstate we have:

$$Y_{\ell,\ell}(\theta,\phi) = \left[\frac{(2\ell+1)!}{4\pi} \right]^{\frac{1}{2}} \frac{\sin^\ell \theta}{2^\ell \ell!} e^{i\ell\theta}. \qquad (6.218)$$

The next step is to obtain the spherical harmonic expressions when $m \neq \ell$. This can be achieved either by repeated applications of the lowering operator L_- to the eigenstates $Y_{\ell,\ell}(\theta,\phi)$ of eq. 6.218 or by repeated applications of the raising operator L_+ to the lowest state $Y_{\ell,-\ell}(\theta,\phi)$.

Let's study the application of the operator L^2 to the eigenstates resulting from the application of $L_\pm$ to a spherical harmonic $Y_{\ell,m}(\theta,\phi)$. Because L^2 commutes[22] with $L_\pm$ and using 6.203, we have:

$$L^2 \left(L_\pm Y_{\ell,m} \right) = L_\pm L^2 Y_{\ell,m} = \ell(\ell+1)\hbar^2 \left(L_\pm Y_{\ell,m} \right). \qquad (6.219)$$

Because L^2 commutes with L_z, it is also true that:

$$L^2 \left(L_z Y_{\ell,m} \right) = L_z L^2 Y_{\ell,m} = \ell(\ell+1)\hbar^2 \left(L_z Y_{\ell,m} \right). \qquad (6.220)$$

---

[22]L^2 commutes with L_x and L_y and thus with any linear combinations of L_x and L_y.

This means that applying the operators $L_\pm$ and L_z to $Y_{\ell,m}$ does not change the value of ℓ and the new eigenstate has the same eigenvalue $\ell(\ell+1)\hbar^2$.

On the other hand, the eigenvalue of L_z is changed when operating with $L_\pm$. In fact:

$$
\begin{aligned}
L_z\left(L_\pm Y_{\ell,m}\right) &= L_\pm L_z Y_{\ell,m} + [L_z, L_\pm] Y_{\ell,m} \\
&= m\hbar(L_\pm Y_{\ell,m}) \pm \hbar L_\pm Y_{\ell,m} \\
&= (m\pm 1)\hbar(L_\pm Y_{\ell,m})
\end{aligned}
\tag{6.221}
$$

where we have used:

$$
[L_\pm, L_z] = [L_x, L_z] \pm i[L_y, L_z] = i\hbar(-L_y \pm iL_x) = \mp\hbar L_\pm.
\tag{6.222}
$$

Eq. 6.221 shows that $(L_\pm Y_{\ell,m})$ is an eigenstate of L_z We can therefore write, in general, that:

$$
L_\pm Y_{\ell,m} = c_\pm(\ell,m)Y_{\ell,m\pm 1}
\tag{6.223}
$$

showing that $L_\pm$ raise/lower the eigenvalues of L_z by $\hbar$ while keeping the eigenvalues of L^2, i.e. $(\ell(\ell+1)\hbar)$, unaltered. Repeated applications of the lowering operator L_- to $Y_{\ell,m}$ must stop when we reach the lower limit of $m = -\ell$, i.e.:

$$
L_- Y_{\ell,-\ell} = 0.
\tag{6.224}
$$

Analogously, we have that repeated applications of the raising operator must stop when we reach the upper limit $m = \ell$, i.e.:

$$
L_+ Y_{\ell,\ell} = 0.
\tag{6.225}
$$

Next, we compute the coefficients $c_\pm(\ell,m)$ of eq. 6.223. Using eq. 6.223, we first compute the norm:

$$
\begin{aligned}
\langle L_\pm Y_{\ell,m} | L_\pm Y_{\ell,m} \rangle &= |c_\pm(\ell,m)|^2 \langle Y_{\ell,m\pm 1} | Y_{\ell,m\pm 1} \rangle \\
&= |c_\pm(\ell,m)|^2.
\end{aligned}
\tag{6.226}
$$

The same norm can be computed by:

$$
\begin{aligned}
\langle L_\pm Y_{\ell,m} | L_\pm Y_{\ell,m} \rangle &= \langle Y_{\ell,m} | L_\mp L_p m Y_{\ell,m} \rangle \\
&= \langle Y_{\ell,m} | (L^2 - L_z^2 \mp \hbar L_z) Y_{\ell,m} \rangle \\
&= \langle Y_{\ell,m} | L^2 Y_{\ell,m} \rangle - \langle Y_{\ell,m} | L_z^2 Y_{\ell,m} \rangle \mp \hbar \langle Y_{\ell,m} | L_z Y_{\ell,m} \rangle \\
&= (\ell(\ell+1) - m^2 \mp m)\hbar^2.
\end{aligned}
\tag{6.227}
$$

Equating eqs. 6.226 and 6.227, we find an expression for the coefficients $c_\pm$:

$$
c_\pm(\ell,m) = \hbar\sqrt{\ell(\ell+1) - m(m\pm 1)}.
\tag{6.228}
$$

Therefore we have that the raising/lowering operators acting on the spherical harmonics obey the following eigenvalue equation:

$$
L_\pm Y_{\ell,m} = \hbar\sqrt{\ell(\ell+1) - m(m\pm 1)}Y_{\ell,m\pm 1}.
\tag{6.229}
$$

Eq. 6.229 can be inverted to give:

$$Y_{\ell,m\pm 1} = \frac{1}{\hbar \sqrt{\ell(\ell+1) - m(m\pm 1)}} Y_{\ell,m}. \tag{6.230}$$

Eq. 6.230 defines $Y_{\ell,m}(\theta,\phi)$ recursively.

The spherical harmonics are usually written in terms of so called **associated Legendre functions** $P_{\ell,m}(\cos\theta)$ which in turn are obtained by the **Legendre polynomials**:

$$P_\ell(x) = \frac{1}{2^\ell \ell!} \left(\frac{d}{dx}\right)^\ell (x^2 - 1)^\ell. \tag{6.231}$$

The associated Legendre functions are then written as:

$$P_{\ell,m}(x) = (-1)^m (1-x^2)^{\frac{m}{2}} \frac{d^m}{dx^m}(P_\ell(x)). \tag{6.232}$$

We can finally give the expression for the spherical harmonics as:

$$Y_{\ell,m}(\theta,\phi) = (-1)^m \sqrt{\frac{2\ell+1}{4\pi}\frac{(\ell-m)!}{(\ell+m)!}} e^{im\phi} P_{\ell,m}(\cos\theta). \tag{6.233}$$

6.2.7 SPIN

The lengthy arguments of the previous section have shown that we are justified to split the total angular momentum $J = L + S$ into two separate components L and S, where L is the orbital angular momentum and can assume only integer multiples of $\hbar$ while S is the intrinsic angular momentum and it can assume integer and half-integer multiples of $\hbar$.

We now go back to the Stern and Gerlach experiment described in section 6.2.4. We have seen that such an experiment had the unexpected result of showing that the most distant electron in a silver atom was responsible of splitting the beam into two components. Assuming that the orbital angular moment was zero (we'll go back later on this assumption), the splitting was due to an intrinsic magnetic moment of the electron. One hypothesis consisted in treating the electron as a rotating charge producing a magnetic moment. Having only two orientations for the rotation around its axis, that might justify the two values for the intrinsic magnetic moment of $\pm\frac{\hbar}{2}$. Thus the name **spin**.

When we have a spinning mass we know, from classical physics, that there is an associated angular momentum. If the mass is electrically charged, then we have also an associated magnetic moment.

Let us, for a moment, describe the orbiting electron in an atom as a particle of mass $m_e = 9.1 \times 10^{-31}$ kg and charge[23] $e = -1.6 \times 10^{-19}C$ as depicted in fig. 6.6 as

---

[23]The charge of 1 Coulomb (C) is the amount of charge delivered by a DC current of 1 Ampere (A) flowing in 1 second and corresponds to approximately $6.2 \times 10^{18} e$.

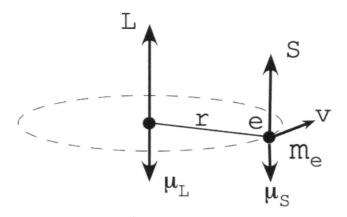

Figure 6.6 Angular and magnetic moments of an orbiting electron.

described in section 6.2.4. This is obviously a classical picture that we know is not accurate: in quantum mechanics the electron does not orbit around the nucleus but it is described by a complex wavefunction ψ spreading over the entire space. We can only calculate the probability that an electron is at a certain position in space at a given time by calculating the square modulus of ψ. However, let's keep this picture for a moment so that we can estimate the classical angular and magnetic moment and show that they are proportional to each other and the proportionality factor is a universal constant[24].

From eq. 6.124 we see that the ratio, called *gyromagnetic ratio*:

$$\gamma = \frac{\mu_L}{L} = \frac{e}{2m_e} \tag{6.234}$$

does not depend on the radius of the orbit nor the velocity of the orbiting electron. It is therefore natural to identify the anomalous split observed by Stern and Gerlach as due to an intrinsic magnetic moment produce by assuming that the electron is a small rigid charge spinning around an axis with an angular momentum S analogous to L. The simplest assumption would be to assume that the electron possesses an intrinsic magnetic moment μ_S whose ratio is proportional to the value in eq. 6.234 but with an intrinsic angular momentum S instead of L:

$$\gamma_e = \frac{\mu_S}{S} = g\frac{e}{2m_e} \tag{6.235}$$

where g is a proportionality factor called **Lande' factor** or **g-factor**. Experiments, and later a relativistic theory of electrodynamics, have shown that the g-factor is $g = 2$.

---

[24]A universal constant is a physical constant assumed to be the same value everywhere in the Universe and not changing with time.

We can rewrite eq. 6.235 in vector form as:

$$\vec{\mu_S} = -g\mu_B \frac{\vec{S}}{\hbar} \tag{6.236}$$

where $\mu_B = \frac{e\hbar}{2m_e}$ is Bohr magneton. The minus sign in eq. 6.236 is due to the convention that the electron has a negative charge thus explaining why μ_L, μ_S points opposite to L, S in fig. 6.6.

We need to point out immediately, as Wolfgang Pauli did, that the model of a rotating charge for the electron spin is not correct, at least in its simple version[25]. In fact, if we assume that the electron is a sphere of radius equal to the so called *classical electron radius* $r_e \approx 10^{-15}$ m, mass m_e with a distributed charge e, the rotation at its equator would be $v = \omega r$, where ω is the angular velocity. Assuming a moment of inertia $I = \frac{2}{5}m_e r_e^2$, the angular momentum is $\omega = \frac{v}{r} = \frac{S}{I} = \frac{\hbar}{2}\frac{5\hbar}{4m_e r_e^2}$ from which we obtain $v = \frac{\hbar}{2}\frac{5\hbar}{4m_e r_e} \gg c$, where c is the speed of light. Therefore, such a simple model of spin requires the equator of the electron to rotate with a velocity of the order of 200 times the speed of light!

Let's develop a quantum mechanical description of the spin. The Stern and Gerlach experiment has provided a value for the intrinsic angular momentum of the electron that, when measured along the z-axis, is equal to:

$$S_z = \pm\frac{\hbar}{2}. \tag{6.237}$$

This value is consistent with the experimental results once the geometry of the Stern and Gerlach experiment together with the strength of the external magnetic field is taken into account. A particle whose the measurements of their intrinsic angular momentum is equl to $\pm\frac{\hbar}{2}$ is called a **spin one-half** particle. The associated intrinsic magnetic moment is one Boht magneton as shown in eq. 6.236.

The Stern and Gerlach apparatus (SG from now on) can be thought of as a device that takes as input a beam of particles and it separates it into 2-beams (for spin one-half) parallel and anti-parallel to the direction $\vec{n} = (x, y, z)$ of the external magnetic field gradient (see fig. 6.7). Since there is no preferred direction, the usual convention is to align the external magnetic field gradient along the z-direction.

An electron (or an Ag atom) therefore needs to be described by a wavefunction that depends on the position r, the time t and the spin S_z:

$$\psi = \psi(r, t, S_z) \tag{6.238}$$

where the norm $|\psi(r, t, S_z)|^2 d^3r$ represents the probability of finding the electron within the volume d^3r around r at the time t and with spin S_z. However, while the coordinates r and t are continuous, the spin coordinate is discrete and, in the case

---

[25]For some attempts at reviving the spinning electron model see, for example Ohanian [32] or Sebens [39].

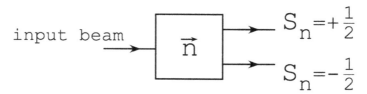

Figure 6.7 Schematic representation of a Stern and Gerlach experiment. An input beam of spin one-half particles in input is divided into two beams along the direction $\vec{n}$: the upper beam contains spin $+\frac{1}{2}$ only particles while the lower beam contains spin $-\frac{1}{2}$ only particles.

of spin one-half, is restricted to two possible values $S_z = \pm\frac{\hbar}{2}$. Therefore it comes natural, and convenient, to write the wave function 6.238 as a combination of two wavefunctions:

$$\psi_+(r,t) = \psi(r,t,+\frac{\hbar}{2})$$
$$\psi_-(r,t) = \psi(r,t,-\frac{\hbar}{2}). \tag{6.239}$$

We can make a further assumption and consider the two wavefunctions of eq. 6.239 as the two components of a vector depending only on r and t and the spin dependence is represented simply by which component of the vector is considered:

$$\psi = \begin{pmatrix} \psi_+ \\ \psi_- \end{pmatrix} \tag{6.240}$$

where we have dropped the (r,t) dependence for simplicity. The special 2-dimensional vector in eq. 6.240 is called **Spinor**. Our mathematical model of spin one-half therefore consists of the 3-dimensional space where, at each coordinate (r,t) there is an associated two-component spinor. If we are only interested in studying the spin, we can define the two kets:

$$|z,0\rangle = \psi_+$$
$$|z,1\rangle = \psi_- \tag{6.241}$$

where $|z,0\rangle$ represents spin "up" with respect to the z-direction while the ket $|z,1\rangle$ represents spin "down" with respect to the z-direction. If we were to align the external magnetic field gradient in the SG experiment along the x or y directions, then the corresponding kets would be, respectively, $|x,0\rangle, |y,0\rangle$ for spin up and $|x,1\rangle, |y,1\rangle$ for spin down.

Let's now imagine a series of experiments or, as physicists like to call it, *thought experiments*[26], where we position a certain number of Stern and Gerlach apparatuses in series. This thought experiment will allow us to determine the form of the operators capable of acting on the spinors.

---

[26] A thought experiment in physics, sometimes called *gedankenexperiment* in German, is a potentially doable experiment used to prove a theory or a principle without actually performing it in practice.

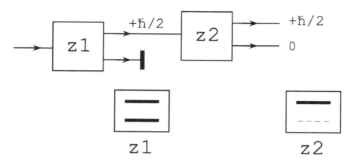

Figure 6.8 Two Stern and Gerlach (SG) apparatuses in series. A beam of spin one-half particles (spin up) is filtered by the first SG. The filtered beam then enters a second SG apparatus showing that $S_z = +\frac{\hbar}{2}$ particles have no $S_z = -\frac{\hbar}{2}$ component.

Fig. 6.8 shows two SG apparatuses both having the external magnetic field gradient oriented along the z-direction which we indicate with SGz. The "z" reminds us of the magnetic field gradient orientation. A beam of spin one-half particles enters the first SGz separating the beam into an upper beam "spin up" ($S_z = +\frac{\hbar}{2}$) and a lower beam "spin down" ($S_z = -\frac{\hbar}{2}$) as shown on the screen z1. The spin down beam is blocked and does not propagate further. The emerging spin up beam then enters a second SGz which, acting already on a spin up beam outputs only spin up particles as shown in the screen z2.

This first thought experiment is simply telling us that the two states 6.241 are to be considered as basis eigenstates: in fact, the second filtering of SGz shows that $S_z = +\frac{\hbar}{2}$ states do not contain $S_z = -\frac{\hbar}{2}$ states meaning that they are orthogonal. In addition, all states spin up entering the first SGz apparatus emerge entirely at the output of the second SGz apparatus indicating that the two states are parallel. We must therefore have that:

$$\langle z, 0 | z, 1 \rangle = 0$$
$$\langle z, 1 | z, 0 \rangle = 0$$
$$\langle z, 0 | z, 0 \rangle = 1$$
$$\langle z, 1 | z, 1 \rangle = 1. \tag{6.242}$$

We can therefore write the eigenvalue equations:

$$S_z | z, 0 \rangle = +\frac{\hbar}{2} | z, 0 \rangle$$
$$S_z | z, 1 \rangle = -\frac{\hbar}{2} | z, 1 \rangle . \tag{6.243}$$

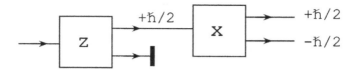

Figure 6.9 Two Stern and Gerlach (SG) apparatuses in series. A beam of spin one-half particles (spin up) is filtered by the first SG oriented along the z-direction. The filtered beam then enters a second SG apparatus oriented along the x-direction showing that the beam is separated into two beams corresponding to spin up (along x) and spin down (along x). Classically there should not be any split beam from the second SG.

The following two-components vectors are a perfectly valid representation of the two eigenbasis $\langle z, 0|, \langle z, 1|$:

$$|z, 0\rangle \rightarrow \begin{pmatrix} 1 \\ 0 \end{pmatrix}$$
$$|z, 1\rangle \rightarrow \begin{pmatrix} 0 \\ 1 \end{pmatrix}$$

(6.244)

In fact, they satisfy the relations 6.242 thus forming an orthonormal basis. According to our definitions, any spin state can be represented as a linear combination of the two basis 6.244:

$$|\psi\rangle = c_0 \begin{pmatrix} 1 \\ 0 \end{pmatrix} + c_1 \begin{pmatrix} 0 \\ 1 \end{pmatrix} = \begin{pmatrix} c_0 \\ c_1 \end{pmatrix}$$

(6.245)

where c_0, c_1 are two complex numbers.

The operator S_z acting on 2-components vectors like 6.245 must therefore be a 2×2 hermitian matrix. In fact, it is easy to verify that the operator S_z can be written as a 2×2 matrix of the form:

$$S_z = \frac{\hbar}{2}\sigma_z = \frac{\hbar}{2}\begin{pmatrix} 1 & 0 \\ 0 & -1 \end{pmatrix}.$$

(6.246)

where $\sigma_z = \begin{pmatrix} 1 & 0 \\ 0 & -1 \end{pmatrix}$ is a Hermitian 2×2 matrix. Notice that S_z verifies the eigenvalues equations 6.243.

Let's now discuss the series of Stern Gerlach apparatuses shown in fig. 6.9. A first SGz filters out a $+\frac{\hbar}{2}$ spin up beam. Notice that we have a spin up with respect to the z-direction. This filtered beam is directed toward the input of a second SGx, i.e. a Stern and Gerlach apparatus with the gradient of the magnetic field now oriented along the x-direction, orthogonal to the z-direction. Classically we would expect no splitting out of the second SG because we have already selected a beam of spins all oriented perpendicularly to x. Instead, nature shows us that, out of the SGx apparatus, we have the beam split into two beams $+\frac{\hbar}{2}$ and $-\frac{\hbar}{2}$ with equal intensity, i.e. with 50% probability of spin up and spin down along x. This means that, in quantum mechanics, a state with $S_z = \frac{\hbar}{2}$ has components along the orthogonal direction x equal to $\pm\frac{\hbar}{2}$. Mathematically we can describe this by writing that the state $|z,0\rangle$ of spin-up in the z-direction can be written as a linear combination of the states $|x,o\rangle$ and $|x,1\rangle$ with equal coefficients:

$$|z,0\rangle = \frac{1}{\sqrt{2}}|x,0\rangle + \frac{1}{\sqrt{2}}|x,1\rangle. \tag{6.247}$$

It is easy to verify that the orthogonal state $|z,1\rangle$ must have the form:

$$|z,1\rangle = \frac{1}{\sqrt{2}}|x,0\rangle - \frac{1}{\sqrt{2}}|x,1\rangle. \tag{6.248}$$

We are now after the matrix representation for the operator $S_x = \frac{\hbar}{2}\sigma_x$, where σ_x is a 2×2 matrix like σ_z.

The operator σ_x has two eigenvalues $\lambda_0 = +1, \lambda_1 = -1$ and, because of eqs. 6.247 and 6.248, it has normalized eigenvectors:

$$u_0 = \frac{1}{\sqrt{2}}\begin{pmatrix}1\\1\end{pmatrix}$$

$$u_1 = \frac{1}{\sqrt{2}}\begin{pmatrix}1\\-1\end{pmatrix}. \tag{6.249}$$

We can determine the matrix σ_x given its eigenvectors and eigenvalue in the following way: we first build a diagonal matrix D where, on its diagonal, we put the eigenvalues:

$$D = \begin{pmatrix}1 & 0\\0 & -1\end{pmatrix} \tag{6.250}$$

then we build the matrix M made inserting the corresponding eigenvectors as columns:

$$M = \frac{1}{\sqrt{2}}\begin{pmatrix}1 & 1\\1 & -1\end{pmatrix}. \tag{6.251}$$

We have:

$$\sigma_x = MDM^{-1} = \begin{pmatrix}0 & 1\\1 & 0\end{pmatrix}. \tag{6.252}$$

Finally, under the assumption that the matrices S or σ are representing the intrinsic angular momentum, they must obey the same commutation relations obeyed by the orbital angular momentum L. Therefore, we find $S_y = \frac{\hbar}{2}\sigma_y$:

$$i\hbar S_y = [S_z, S_x] \tag{6.253}$$

from which we obtain:

$$\sigma_y = \begin{pmatrix}0 & -i\\i & 0\end{pmatrix}. \tag{6.254}$$

In summary, we have that:

$$S_x = \sigma_x, \quad S_y = \sigma_y, \quad S_z = \sigma_z \tag{6.255}$$

where:

$$\sigma_x = \begin{pmatrix} 0 & 1 \\ 1 & 0 \end{pmatrix}, \quad \sigma_y = \begin{pmatrix} 0 & -i \\ i & 0 \end{pmatrix}, \quad \sigma_z = \begin{pmatrix} 1 & 0 \\ 0 & -1 \end{pmatrix}. \tag{6.256}$$

The three matrices above, called **Pauli matrices** already discussed in eq. 6.18, together with the 2×2 identity matrix form an orthonormal basis for all 2×2 matrices. They all have eigenvalues equal to ± 1 and obey the commutation relations:

$$[\sigma_i, \sigma_j] = 2i\varepsilon_{ijk}\sigma_k \tag{6.257}$$

where we numbered the Pauli matrices according to $\sigma_x = \sigma_1$, $\sigma_y = \sigma_2$ and $\sigma_z = \sigma_3$.

In complete analogy with the orbital angular momentum L, we can build the following operators defined in the tridimensional space:

$$S = S_x e_x + S_y e_y + S_z e_z$$
$$S^2 = S_x^2 + S_y^2 + S_z^2. \tag{6.258}$$

The spin operators satisfy the following commutation relations:

$$[S_x, S_y] = i\hbar S_z$$
$$[S_y, S_z] = i\hbar S_x$$
$$[S_z, S_x] = i\hbar S_y \tag{6.259}$$
$$[S^2, S_i] = 0, \quad \text{for } i = x, y, z.$$

Continuing with the analogy, while an eigenstate of L are indicated with the ket $|\ell, m\rangle$, the eigenstates of spin S are indicated with $|s, m\rangle$ where s is half-integer and m has $(2s+1)$ values. The spin eigenvalues equation equivalent to eqs. 6.169 are:

$$S^2 |s, m\rangle = s(s+1)\hbar^2 |s, m\rangle$$
$$S_z |s, m\rangle = m\hbar |s, m\rangle \tag{6.260}$$

where s is half-integer and m is bound to $(2s+1)$ values $-s, -s+1, ..., s$. For a spin $s = \frac{1}{2}$ state, $m = \pm \frac{1}{2}$.

The spin is an intrinsic property of elementary particles like, for example, the electron and cannot be changed. Particles with half-integer spin are called **fermions** while particles with integer spin are called **bosons**.

We have already calculated the matrix form of the S_z operator for a spin $\frac{1}{2}$ particle:

$$S_z = \frac{\hbar}{2}\sigma_z = \frac{\hbar}{2}\begin{pmatrix} 1 & 0 \\ 0 & -1 \end{pmatrix}. \tag{6.261}$$

The matrix form of the S^2 operator can be calculated using the formula for matrix elements 6.46. Given the two base vectors $|0\rangle$ and $|1\rangle$, we know that the S^2 operator is a 2×2 matrix of the form:

$$S^2 = \begin{pmatrix} S_{11}^2 & S_{12}^2 \\ S_{21}^2 & S_{22}^2 \end{pmatrix}. \tag{6.262}$$

Using eq. 6.46, we have:

$$S_{11}^2 = \langle 0|S^2|0\rangle = \langle 0|s(s+1)\hbar^2|0\rangle = \frac{3}{4}\hbar^2\langle 0|0\rangle = \frac{3}{4}\hbar^2$$

$$S_{12} = \langle 0|S^2|1\rangle = \frac{3}{4}\hbar^2\langle 0|1\rangle = 0$$

$$S_{21} = \langle 1|S^2|0\rangle = \frac{3}{4}\hbar^2\langle 1|0\rangle = 0 \tag{6.263}$$

$$S_{22} = \langle 1|S^2|1\rangle = \frac{3}{4}\hbar^2\langle 1|1\rangle = \frac{3}{4}\hbar^2$$

and the S^2 matrix can be written as:

$$S^2 = \frac{3}{4}\hbar^2 \begin{pmatrix} 1 & 0 \\ 0 & 1 \end{pmatrix}. \tag{6.264}$$

Notice the similarity of S^2 and S_z with the expressions of J^2 and J_z, respectively, in eqs. 6.136 and 6.137. Since the ket $|J\rangle$ has 3 components, then the angular momentum operators are represented by 3×3 matrices as opposed to the 2 component spinors with corresponding 2×2 matrices.

6.3 THE HYDROGEN ATOM

We have studied in the previous sections that the angular momentum operator J can be described as the combination of two angular momenta according to $J = L + S$ where L is the orbital angular momentum and S is the intrinsic angular momentum, or spin. We have seen that it is more convenient to study the square of the orbital angular momentum, L^2, and the projection L_z along an (arbitrary) axis z. These two operators commute and therefore they admit a common set of eigenstates characterized by two quantum numbers ℓ and m. The quantum number ℓ determines the magnitude of the angular momentum while the quantum number m determines the projection of the angular momentum along the z-axis.

In the case of the orbital angular momentum, ℓ can assume only integer values and m is restricted to the range $m = -\ell, -\ell+1, ..., +\ell$. The intrinsic angular momentum, or spin, can assume integer of half-integer values.

Associated with the spin, and in analogy with the orbital angular momentum, we have an additional quantum number, analogous to m, which we indicate as m_s which is subject to the same restriction as m, i.e. $m_s = -\ell, -\ell+1, ..., +\ell$ which indicates the projection of the spin along the z-direction. In the special case of spin one-half, the quantum number m_s can assume only two values $m_s = +\frac{1}{2}$ and $m_s = -\frac{1}{2}$

corresponding to the somehow arbitrary definition of "spin up" and "spin down" and responsible for the splitting of the beam of Ag atoms in an inhomogeneous magnetic field in the Stern and Gerlach experiment.

The next logical step would consist in study the Hydrogen atom which is composed by a positively charged proton (charge $+e$) and a negatively charged electron (charge $-e$). The resulting potential is the so called *Coulomb potential* given by:

$$V(r) = -\frac{e^2}{4\pi\varepsilon_0} \tag{6.265}$$

where r is the separation between the proton and the electron. Classically, the proton and the electron orbit around the center of mass of the two-body system. However, considering that the mass of the electron is much smaller than the mass of the proton[27], we can assume that the proton sits in the center of the coordinate system and the electron orbits the proton.

In our treatment, we make the simplifying assumptions that the electron and the proton do not have spin and that the particles move non relativistically, i.e. the typical velocities involved are $v \ll c$, where c is the speed of light.

The Hamiltonian of the hydrogen atom in spherical coordinates is:

$$H(r,\theta,\phi) = -\frac{\hbar^2}{2mr^2}\left[\frac{\partial}{\partial r}\left(r^2\frac{\partial}{\partial r}\right) + \frac{1}{\sin\theta}\frac{\partial}{\partial\theta}\left(\sin\theta\frac{\partial}{\partial\theta}\right) + \frac{1}{\sin^2\theta}\frac{\partial^2}{\partial\phi^2}\right]$$
$$-\frac{e^2}{4\pi\varepsilon_0 r} \tag{6.266}$$

where m is the electron mass. The Hamiltonian 6.266 allows us to write the time independent Schrödinger equation for the hydrogen atom as:

$$\left[-\frac{\hbar^2}{2mr^2}\left[\frac{\partial}{\partial r}\left(r^2\frac{\partial}{\partial r}\right) + \frac{1}{\sin\theta}\frac{\partial}{\partial\theta}\left(\sin\theta\frac{\partial}{\partial\theta}\right) + \frac{1}{\sin^2\theta}\frac{\partial^2}{\partial\phi^2}\right] - \frac{e^2}{4\pi\varepsilon_0 r}\right]\psi = E\psi \tag{6.267}$$

where the wavefunction $\psi = \psi(r,\theta,\phi)$ depends on the radial coordinate r and the spherical angles θ and ϕ.

Using the expression in spherical coordinates of the operator L^2 from eq. 6.189, we can write eq. 6.267 in a more compact form:

$$\left[-\frac{\hbar^2}{2m}\left(\frac{1}{r^2}\frac{\partial}{\partial r}r^2\frac{\partial}{\partial r} - \frac{L^2}{\hbar^2 r^2}\right) - \frac{e^2}{4\pi\varepsilon_0 r}\right]\psi = E\psi. \tag{6.268}$$

After a little algebra, we find that eq. 6.268 becomes:

$$\hbar^2\frac{\partial}{\partial r}\left(r^2\frac{\partial}{\partial r}\psi\right) + 2mr^2(E-V)\psi = L^2\psi. \tag{6.269}$$

---

[27] The mass of the proton is about 1,836 times the mass of the electron.

We have seen that the angular momentum operator, with its spherical harmonics eigenfunctions $Y_{\ell,m}(\theta,\phi)$, does not depend on the radial coordinate r. This means that the radial dependence is limited to the left-hand side of eq. 6.269. This allows us to separate variables by assuming that:

$$\psi(r,\theta,\phi) = R(r)Y_{\ell,m}(\theta,\phi). \tag{6.270}$$

If we divide each side of eq. 6.269 by the $\psi(r,\theta,\phi)$ of eq. 6.270 we obtain:

$$\frac{\hbar^2}{R(r)}\frac{\partial}{\partial r}r^2\frac{\partial}{\partial r}R(r) + \frac{2mr^2}{R(r)}[E-V]R(r) = \frac{1}{Y_{\ell,m}(\theta,\phi)}L^2 Y_{\ell,m}(\theta,\phi). \tag{6.271}$$

The left-hand side of eq. 6.271 depends only on the radial coordinate r while the right-hand side depends only on the angular coordinates θ,ϕ. Therefore we can separate eq. 6.271 into two equations:

$$\frac{\hbar^2}{R(r)}\frac{\partial}{\partial r}r^2\frac{\partial}{\partial r}R(r) + \frac{2mr^2}{R(r)}[E-V]R(r) = \lambda$$

$$\frac{1}{Y_{\ell,m}(\theta,\phi)}L^2 Y_{\ell,m}(\theta,\phi) = \lambda \tag{6.272}$$

where λ does not depend on (r,θ,ϕ). The second equation in 6.272 is the eigenvalue equation for the angular momentum operator L^2 that we have already studied (see, for example, the first equation in 6.203). We know therefore that the constant is:

$$\lambda = \ell(\ell+1)\hbar^2. \tag{6.273}$$

Inserting eq. 6.273 into the first eq. 6.271, and after some algebra, we obtain the so called Schrödinger radial equation for the hydrogen atom:

$$\left[-\frac{\hbar^2}{2mr^2}\frac{\partial}{\partial r}r^2\frac{\partial}{\partial r} + \frac{\ell(\ell+1)\hbar^2}{2m}\frac{1}{r^2} - \frac{1}{4\pi\varepsilon_0}\frac{e^2}{r}\right]R(r) = ER(r). \tag{6.274}$$

The separation of variables 6.270 has allowed us to write the radial Schrödinger equation in the form 6.274 where to the Coulomb potential energy it is added an additional term leading to an *effective* potential energy of the form:

$$V_{eff} = V_{cen} + V_{Cou} = \frac{\ell(\ell+1)\hbar^2}{2m}\frac{1}{r^2} - \frac{e^2}{4\pi\varepsilon_0}\frac{1}{r}. \tag{6.275}$$

The effective potential energy 6.275 contains, in addition to the standard attractive Coulomb potential V_{Cou} proportional to $1/r$, a *centrifugal* repulsive term V_{cen} with opposite sign proportional to $1/r^2$. The Coulomb potential is responsible of an attractive force $F = -\frac{dV_{Cou}}{dr} < 0$ directed from the electron to the proton due to the fact that the derivative of the potential is always positive. The centrifugal force $F = -\frac{dV_{cen}}{dr} > 0$, on the other hand, is always positive being the derivative of the potential energy which is always negative. In fig. 6.10 an example of an effective

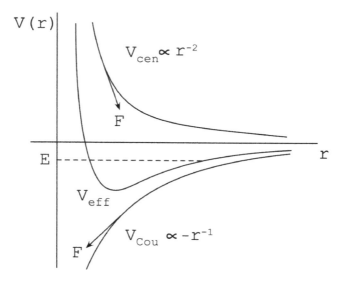

Figure 6.10 Potential energy of the hydrogen atom showing the repulsive centrifugal term proportional to r^{-2} combined with the Coulomb term proportional to $-r^{-1}$ to give the effective potential.

potential corresponding to a particular value of the quantum number ℓ is shown. Notice that, in the special case of $\ell = 0$, the centrifugal force is zero and the electron is subject only to the standard Coulomb potential.

In general, the shape of V_{eff} depends on ℓ which determines the strength of the centrifugal force: higher values of ℓ correspond to stronger centrifugal forces and the position of the minimum of the V_{eff} potential is shifted to higher values of the radius r. As in the classical case of a two body problem, positive values of the total energy E correspond to free particles while negative values of E correspond to bound states.

Let's introduce a new variable:

$$u(r) = rR(r). \tag{6.276}$$

Eq. 6.274 depends only on the radial coordinate r and therefore, after changing to the new variable u defined in 6.276, it can be written in terms of total derivatives as:

$$\frac{d^2u}{dr^2} + \frac{2m}{\hbar^2}\left[E + \frac{e^2}{4\pi\varepsilon_0}\frac{1}{r} - \frac{\ell(\ell+1)\hbar^2}{2mr^2}\right]u = 0. \tag{6.277}$$

Eq. 6.277 can be solved using various techniques like, for example, Frobenius[28] method. We will discuss here a different way to find the solution. We start by studying

---

[28]Ferdinand Georg Frobenius (1849-1917) was a German mathematician known for his contributions to the theory of differential equations. He has made important contributions to group theory and number theory.

the solution in the two limiting cases when $r \to \infty$ and and when $r \to 0$. When r is very large, eq. 6.277 simplifies to:

$$\frac{d^2u}{dr^2} + \frac{2m}{\hbar^2}Eu = 0 \tag{6.278}$$

which admits as a solution:

$$u = A_1 e^{-\sqrt{\frac{2mE}{\hbar^2}}r} + A_2 e^{\sqrt{\frac{2mE}{\hbar^2}}r}. \tag{6.279}$$

As a boundary condition, the wavefunction must tend to zero at infinity and therefore we must have $A_2 = 0$.

In the limiting case of small r, eq. 6.277 simplifies to:

$$\frac{d^2u}{dr^2} - \frac{\ell(\ell+1)}{r^2}u = 0 \tag{6.280}$$

which admits two independent solutions:

$$u_1 = B_1 r^{\ell+1}$$
$$u_2 = B_2 \frac{1}{r^\ell}. \tag{6.281}$$

Again, as a boundary condition we need $B_2 = 0$ to keep the wavefunction finite when $r \to 0$. The general solution can be written as:

$$u = Ce^{-\sqrt{\frac{2mE}{\hbar^2}}r}f(r)r^{\ell+1} \tag{6.282}$$

where $f(r)$ is still to be determined. Let's assume that the function $f(r)$ is represented by a series expansion of the form:

$$f(r) = \sum_{k=0}^{\infty} a_k r^k. \tag{6.283}$$

If we insert the guess function 6.283 into eq. 6.277, we obtain a recursion relation for the coefficients a_k of the form:

$$a_k = -2a_{k-1} \frac{\frac{me^2}{4\pi\varepsilon_0\hbar^2} - (\ell+k)\sqrt{-\frac{2mE}{\hbar^2}}}{(\ell+k)(\ell+k+1) - \ell(\ell+1)}. \tag{6.284}$$

The series 6.283 with the coefficients 6.284 diverges when $r \to \infty$ unless the there is a value of k for which the numerator of the right-hand side of 6.284 is zero. This is equivalent to require that the function $f(r)$ is a polynomial with a finite number of terms. Therefore, in order to have a well-behaved wavefunction in the interval $0 \le r \le \infty$ we must have:

$$\frac{me^2}{4\pi\varepsilon_0\hbar^2} = (l+k)\sqrt{-\frac{2mE}{\hbar^2}} = n\sqrt{-\frac{2mE}{\hbar^2}} \tag{6.285}$$

where $n = (\ell + k)$ is an integer. Extracting the energy E from eq. 6.285 we obtain:

$$E_n = -\frac{me^4}{32\pi^2 \varepsilon_0^2 n^2 \hbar^2} \tag{6.286}$$

which is exactly Bohr's formula 4.8 obtained with semi-classical arguments.

We have seen that we imposed the condition $n = \ell + k$, where n is called **principal quantum number**. However, it can be shown that the Schrödinger differential equation for the hydrogen atom restricts the range of n to be $n = \ell + 1$ or, equivalently:

$$\ell = 0, 1, ..., (n-1) \tag{6.287}$$

with corresponding polynomials 6.283 in r truncated to the order $n - \ell + 1$.

Let's now find the explicit form of the polynomials $f(r)$ in eq. 6.282. In order to keep the algebra neat, we introduce the variable ρ:

$$\rho = \frac{r}{a_0} \tag{6.288}$$

where $a_0 = \frac{4\pi\varepsilon_0 \hbar^2}{me^2}$ is the Bohr radius making the scale factor ρ dimensionless. Let's introduce two constants:

$$\alpha = \sqrt{\frac{-2mE}{\hbar^2}} \tag{6.289}$$

$$\eta = \alpha a_0.$$

The radial equation can then be written as:

$$\left[\frac{d^2}{d\rho^2} + \frac{2}{\rho} + \frac{\ell(\ell+1)}{\rho^2} - \eta^2 \right] u(\rho) = 0. \tag{6.290}$$

We already know that the wavefunction u must remain finite in the limiting cases of small and large ρ suggesting a solution of the form:

$$u(\rho) = e^{-\eta\rho} \rho^{\ell+1} f(\rho) \tag{6.291}$$

where $f(\rho)$ is the function we are after. Inserting the trial solution 6.291 into 6.290, and after a few re-arrangements, we obtain the following equation for $f(\rho)$:

$$\frac{d^2 f(\rho)}{d\rho^2} + 2\left(\frac{\ell+1}{\rho} - \eta \right) \frac{df(\rho)}{d\rho} + \frac{2}{\rho}(1 - \eta(\ell+1)) f(\rho) = 0. \tag{6.292}$$

We now make the substitution $x = 2\eta\rho$ to obtain:

$$4\eta^2 \frac{d^2 f(\rho)}{d\rho^2} + 4\eta \left(\frac{2\eta(\ell+1)}{\rho} - \eta \right) \frac{df(\rho)}{d\rho} + \frac{4\eta}{x}(1 - \eta(\ell+1)) f(\rho) = 0. \tag{6.293}$$

A further multiplication of eq. 6.293 by $x/4\eta^2$ gives:

$$x\frac{d^2 f(x)}{dx^2} + (2\ell + 2 - x)\frac{df(x)}{dx} + \left(\frac{1}{\eta} - \ell - 1\right)f(x) = 0. \tag{6.294}$$

Eq. 6.294 is equivalent to the **general Laguerre equation** if we make the following substitutions:

$$j = 2\ell + 1$$
$$n = \frac{1}{\eta} \tag{6.295}$$

where $n \geq \ell + 1$. In fact, the general Laguerre equation is:

$$x\frac{d^2 y}{dx^2} + (j + 1 - x)\frac{dy}{dx} + ny = 0. \tag{6.296}$$

Eq. 6.296 admits nonsingular solution only if n is a nonnegative integer. With the substitutions of eq. 6.295, eq. 6.294 becomes:

$$x\frac{d^2 L_{n_r}^j(x)}{dx^2} + (j + 1 - x)\frac{dL_{n_r}^j(x)}{dx} + n_r L_{n_r}^j(x) = 0 \tag{6.297}$$

where $n_r = n - \ell - 1$. We have two quantum numbers: n called **principal quantum number** and n_r called **radial quantum number**. Therefore, quantization in the radial Schrödinger equation is imposed by requesting that its solutions must be nonsingular.

Solution to the generalized Laguerre equation are given by the **generalized Laguerre polynomials** given by:

$$L_{n_r}^j(x) = \frac{e^x x^{-j}}{n_r!}\frac{d^{n_r}}{dx^{n_r}}\left(e^{-x}x^{n_r+j}\right). \tag{6.298}$$

After a bit of algebra, we can finally write down the normalized radial wavefunction as:

$$R_{n\ell}(r) = \sqrt{\left(\frac{2}{na_0}\right)^3 \frac{(n - \ell - 1)!}{2n(n + \ell)!}}\, e^{-\frac{r}{na_0}}\left(\frac{2r}{na_0}\right)^\ell L_{n-\ell-1}^{2\ell+1}(2r/na_0). \tag{6.299}$$

and the full wavefunction of the hydrogen atom is given by:

$$\psi_{n\ell m}(r, \theta, \phi) = R_{n\ell}(r)Y_\ell^m(\theta, \phi). \tag{6.300}$$

Let's summarize the conditions on the three quantum numbers n, ℓ and m:

$$n = 1, 2, 3, \ldots$$
$$\ell = 0, 1, 2, \ldots(n - 1) \tag{6.301}$$
$$m = 0, \pm 1, \pm 2, \ldots, \pm\ell.$$

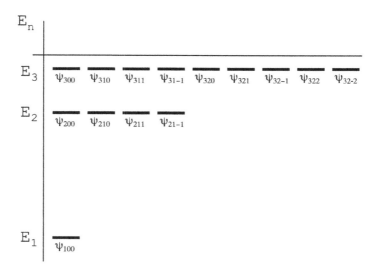

Figure 6.11 The first three degenerate energy levels in the hydrogen atom. Wavefunctions are labeled as $\psi_{n,\ell,m}$.

Physicists call n **principal** quantum number, ℓ **orbital** quantum number and m **magnetic** quantum number. In addition, we have seen that electrons have an additional quantum number m_s called **spin** quantum number.

Although the hydrogen's atom wavefunctions 6.300 depend on the three quantum numbers 6.301, eq. 6.286 tells us that the energy levels are degenerate with respect to ℓ and m meaning that energy levels having the same principal quantum number n, but different quantum numbers ℓ and m, have the same energy as depicted in fig. 6.11.

We are now in a position to write down the complete wavefunctions $\psi_{n,\ell,m}$ for the hydrogen atom. We write down just the first three energy levels $n = 1, 2, 3$.

The first fundamental level corresponding to $n = 1$, $\ell = 0$ and $m = 0$ has the wavefunction:

$$\psi_{100} = \frac{1}{\sqrt{\pi}} \left(\frac{1}{a_0} \right)^{3/2} e^{-\frac{r}{a_0}}. \tag{6.302}$$

The second level $n = 2$ has four degenerate wavefunctions: a single wavefunction when $\ell = 0$, $m = 0$ and three wavefunctions corresponding to $\ell = 1$ and $m = -1, 0, 1$:

$$\psi_{200} = \frac{1}{4\sqrt{\pi}} \left(\frac{1}{a_0} \right)^{3/2} \left(2 - \frac{r}{a_0} \right) e^{-\frac{r}{2a_0}}$$

$$\psi_{210} = \frac{1}{4\sqrt{\pi}} \left(\frac{1}{a_0} \right)^{5/2} r e^{-\frac{r}{2a_0}} \cos\theta \tag{6.303}$$

$$\psi_{21\pm 1} = \frac{1}{8\sqrt{\pi}} \left(\frac{1}{a_0} \right)^{5/2} r e^{-\frac{r}{2a_0}} \sin\theta\, e^{\pm i\phi}.$$

The third level $n = 3$ has nine degenerate wavefunctions: a single wavefunction when $\ell = 0$, $m = 0$; three wavefunctions corresponding to $\ell = 1$ and $m = -1, 0, 1$ and five wavefunctions corresponding to $\ell = 2$ and $m = -2, -1, 0, 1, 2$:

$$\psi_{300} = \frac{1}{81\sqrt{3\pi}} \left(\frac{1}{a_0}\right)^{3/2} \left[27 - \frac{18r}{a_0} + 2\left(\frac{r}{a_0}\right)^2\right] e^{-\frac{r}{3a_0}}$$

$$\psi_{310} = \frac{1}{81}\sqrt{\frac{2}{\pi}} \left(\frac{1}{a_0}\right)^{3/2} \left[\frac{6r}{a_0} - \left(\frac{r}{a_0}\right)^2\right] e^{-\frac{r}{3a_0}} \cos\theta$$

$$\psi_{31\pm1} = \frac{1}{81\sqrt{\pi}} \left(\frac{1}{a_0}\right)^{3/2} \left[\frac{6r}{a_0} - \left(\frac{r}{a_0}\right)^2\right] e^{-\frac{r}{3a_0}} \sin\theta e^{\pm i\theta}$$

$$\psi_{320} = \frac{1}{81\sqrt{6\pi}} \left(\frac{1}{a_0}\right)^{7/2} r^2 e^{-\frac{r}{3a_0}} (3\cos^2\theta - 1) \tag{6.304}$$

$$\psi_{32\pm1} = \frac{1}{81\sqrt{\pi}} \left(\frac{1}{a_0}\right)^{7/2} r^2 e^{-\frac{r}{3a_0}} \sin\theta \cos\theta e^{\pm i\phi}$$

$$\psi_{32\pm2} = \frac{1}{162\sqrt{\pi}} \left(\frac{1}{a_0}\right)^{7/2} r^2 e^{-\frac{r}{3a_0}} \sin^2\theta e^{\pm 2i\phi}.$$

The number of degenerate energy level can be calculated by noticing that for each n there are $(\ell - 1)$ states. In addition, for each ℓ state there are $(2\ell + 1)$ m states. Therefore the degeneracy is:

$$\sum_{\ell}^{n-1} (2\ell + 1) = n^2 \tag{6.305}$$

meaning that there are n^2 degenerate levels with the same n.

In chemistry, the wavefunctions describing a single electron is called **orbital** and since they are 3-dimensional complex functions of space, it is very difficult to construct a visual representation. The interested reader can find several representations of orbitals in the scientific literature.

It is worth mentioning that it is common to find orbitals labeled according to the quantum number ℓ. This *spectroscopic designation* of quantum states has historical origin and is given by:

$$\begin{array}{ccccccc} \ell = & 0 & 1 & 2 & 3 & 4 & 5 \\ & s & p & d & f & g & h. \end{array} \tag{6.306}$$

6.4   PAULI EXCLUSION PRINCIPLE

In the previous section we have seen that solving the Schrödinger equation for the hydrogen atom has shown that the wavefunction describing the orbiting electron depends on three quantum numbers subject to the conditions 6.301. This means that an electron, when encountering a proton, can accommodate itself on a specific orbital starting from the lowest energy. According to eq. 6.286, the orbit with the lowest energy would be the one corresponding to $n = 1$.

If we were to possess the technology to manipulate single particles and we wanted to synthesize more complex atoms, starting from a hydrogen atom, we would be able to make the next atom, the Helium atom, by adding a proton (and two neutrons) to the nucleus. In order to have a neutral atom, we need to add an additional electron. The question is: where such an electron can be accommodated?

If there were no restrictions, then the simplest answer would be to pack as many electrons as possible on the lowest energy state corresponding to $n = 1$, $\ell = 0$ and $m = 0$. Or, in other words, all the electrons on the orbital $1s$ where the 1 indicates the principal quantum number. Unfortunately, this proposal is not corroborated by experimental evidence, including the periodicity of Mendeleev table of elements[29].

We have seen that electrons have half-integer spin which constitutes an additional quantum number. In our ideal assembling of the helium atom, in its ground state, we must include the spin of the electrons. This means that electrons in atoms are characterized by *four* quantum numbers: n, ℓ, m and m_s. The consequences of the spin quantum number, when considering the electronic structure of atoms, i.e. ultimately the whole field of chemistry, are of paramount importance. Let's study the differences in behavior between a beam of hydrogen atoms and a beam of helium atoms when passing through an inhomogeneous magnetic field. We have already studied the Stern and Gerlach experiment where a beam of silver atoms, i.e. atoms with just one electron in an s orbital, is split into two beams revealing the intrinsic magnetic moment, or spin, of the electron. A beam of hydrogen atoms will behave in exactly the same manner showing that hydrogen, like silver, is also split into two beams corresponding to the two quantized states of spin up and spin down with respect to the direction of the inhomogeneous magnetic field.

A beam of helium atoms, in the ground state, when passed through an inhomogeneous magnetic *does not* split into two beams indicating that the addition of another electron into the orbital s has managed to eliminate any magnetic property. We can explain this lack of magnetism by assuming that the two electrons in the $1s$ orbital are *paired* with antiparallel spins, i.e. one electron has spin up and the other has spin down therefore effectively canceling the total spin resulting in zero magnetic moment. It is an experimental fact that states of helium where the two electrons in the $1s$ orbital are both up or both down are not observed. However, if the helium atoms are excited then splitting is observed hinting at the fact that the electron in different orbitals can now have parallel spin.

These experimental facts are hinting at a principle called **Pauli exclusion principle** which states that in a given orbital no two electrons with the same quantum numbers are allowed[30]. Armed with Pauli exclusion principle the periodic table of elements can be built.

---

[29] It is beyond the scope of this book to discuss the beautiful story of the periodic table and its connections with quantum physics.

[30] Wolfgang Pauli (1900-1958) was an Austrian physicist who has contributed to the development of quantum mechanics. In addition to proposing the exclusion principle, for which he received the Nobel prize in 1945, he has hypothesized the existence of the neutrino.

Since the Pauli exclusion principle involves two particles of the same kind, like for example two electrons, we need to study quantum systems of many particles. Classically we can treat many particle systems by labeling each individual particle with an index meaning that we can always follow what the i-th particle does. In quantum mechanics, we know that the dynamics of particles are completely specified by giving their wavefunctions. We have also seen that wavefunctions are often defined over the whole space time. Therefore for example, when dealing with a system composed of two electrons, it will happen that their wavefunctions might overlap. When we detect one of the two electrons the question is: which electron have we detected? Quantum mechanics tells us that we cannot know which of the two electrons we detected mainly because quantum mechanics tells us the *probability* of finding the electron in a certain region of space at a certain time. Quantum mechanics imposes that the two electrons must be described by a single multi-electron wavefunction and there is no way to **distinguish** between the two electrons. In quantum mechanics, electrons are **indistinguishable**.

The fact that electrons are indistinguishable has immediate consequences. Let's indicate with $\psi(r_1, r_2)$ the wavefunction describing two electrons located at the positions r_1 and r_2. If we exchange the position of the two electrons we get a different wavefunction $\psi(r_2, r_1)$. Because the electrons are indistinguishable we must have that the probability of finding an electron in r_1 and an electron in r_2 must not depend on the exchange of electrons. In other terms:

$$|\psi(r_1, r_2)|^2 = \psi^*(r_1, r_2)\psi(r_1, r_2) = \psi^*(r_2, r_1)\psi(r_2, r_1). \tag{6.307}$$

Eq. 6.307 implies that the exchanged wavefunctions satisfy:

$$\psi(r_2, r_1) = e^{i\varphi}\psi(r_1, r_2) \tag{6.308}$$

If we exchange electrons twice we must obtain back the original wavefunction. This means that the original wavefunction has the factor $e^{i\varphi}$ twice. In order to re-obtain the original wavefunction we must have:

$$e^{i\varphi} \cdot e^{i\varphi} = e^{i2\varphi} = 1 \tag{6.309}$$

which means that $\varphi = 0$ or $\varphi = n\pi$ and therefore the phase factor in front of the wavefunction can only assume the values ± 1. The two solutions for φ imply that the wavefunction, under an exchange of particles must obey:

$$\psi(r_1, r_2) = \pm\psi(r_2, r_1). \tag{6.310}$$

Let's introduce the *exchange* operator $\hat{P}$ which, when acting on $\psi(r_1, r_2)$, it exchanges the two particles[31].

---

[31]Considering that, according to quantum mechanics, we cannot distinguish the particles, we should state more correctly that the operator $\hat{P}$ exchanges the unphysical labels, i.e. the labels 1 and 2, in the wavefunction $\psi(r_1, r_2)$.

The operator $\hat{P}$ must leave unchanged the probability amplitude 6.307. Such operator, when applied to a wavefunction, will return $\pm$ the original wavefunction:

$$\hat{P}\psi(r_1, r_2) = \pm\psi(r_2, r_1). \tag{6.311}$$

The eigenvalues of $\hat{P}$ are ± 1 dividing the wavefunctions into two classes, i.e. describing two different kind of particles. We have that particles described by a symmetric wavefunction under $\hat{P}$, i.e. wavefunctions unchanged under exchange of particles, are called **bosons** and have integer spin. Particles anti-symmetric with respect to exchange, i.e. with wavefunctions reversing their sign under exchange of particles, are called **fermions** and have half-integer spin.

It is clear that fermions, being characterized by an anti-symmetric wavefunction, cannot coexist in the same place at the same time. In fact, if the two fermions are in the same location, their wavefunction would be $\psi(r, r)$ which, under exchange would be $-\psi(r, r)$ forcing the wavefunction to be zero. The only escape would be if they have an additional quantum number, the spin, that requires that the two electrons are described by a different wavefunction. Since electrons have spin $1/2$ then a given orbital can accommodate only two electrons providing that they have anti-parallel spin.

Integer spin particles, i.e. *bosons* can co-exist in any number with the same quantum numbers like, for example, the so called **Bose-Einstein condensates**[32] or the photons in a laser cavity.

6.5 HEISENBERG PICTURE

Throughout this book we have represented quantum systems as wavefunctions which are solution to the Schrödinger equation. In this representation of quantum mechanics, the observables are represented by Hermitian operators acting of the wavefunctions. Operators are constant in time and the basis where the operators are defined do change with time. It is a strange representation if we think about our geometric representation of a vector quantity. In our Newtonian mind in fact, vectors change with time, according to Newton's law and the coordinates that we use to represent the vectors are fixed with time. If we indicate with $|\psi_S(t)\rangle$ the time-dependent wavefunction in the Schrödinger representation, then the generic operators are fixed in time. The wavefunctions constitute a basis changing with time according to Schrödinger equation:

$$\frac{d}{dt}|\psi_S(t)\rangle = -\frac{i}{\hbar}\hat{H}|\psi_S(t)\rangle \tag{6.312}$$

---

[32] An example of a Bose-Einstein Condensate is the liquid ^{4}He when below a critical temperature of about 2.17K. ^{4}He atoms have an even number of protons and neutron giving an integer spin and becoming superfluid when below 2.17K. The superfluidity is explained by a partial Bose-Einstein condensation. ^{3}He atoms, i.e. helium atoms where the nucleus have 2 protons and one neutrons, are fermions and do not exhibit superfluidity unless cooled at very low temperatures (2.5 mK) where two atoms can weakly bind and behave as a boson.

which can be formally inverted to give the wavefunction at time t knowing the wave-function at time $t = 0$:

$$|\psi_S(t)\rangle = e^{-i\frac{\hat{H}}{\hbar}t}|\psi_S(0)\rangle \tag{6.313}$$

where the quantum operator Hamiltonian $\hat{H}$ in the exponent of eq. 6.313 is calculated in terms of the power series expansion:

$$e^{-i\frac{\hat{H}}{\hbar}t} = \sum_{n=0}^{\infty} \frac{1}{n!}\left(-i\frac{\hat{H}}{\hbar}t\right)^n. \tag{6.314}$$

Eq. 6.313 can be written as $|\psi_S(t)\rangle = \hat{U}(t)|\psi_S(0)\rangle$, where $\hat{U}(t)$ is the unitary time evolution operator.

We might ask: "Is it possible to have a different representation of quantum mechanics where, in analogy to classical physics, the operators change with time and the basis where the operators are expressed are constant?". The answer is yes and this particular representation is called *Heisenberg picture* of quantum mechanics.

Heisenberg's representation of quantum mechanics, also called *Matrix mechanics* was actually published before Schrödinger mechanics, but it was mathematically more challenging and it took more time to be accepted by the scientific community.

According to the Heisenberg representation of quantum mechanics, the basis $|\psi_H\rangle$ does not change with time. The time dependence is eliminated by formally multiply-ing eq. 6.313 by the factor $e^{i\frac{\hat{H}}{\hbar}t}$:

$$|\psi_H\rangle = e^{i\frac{\hat{H}}{\hbar}t}|\psi_S\rangle = e^{i\frac{\hat{H}}{\hbar}t}e^{-i\frac{\hat{H}}{\hbar}t}|\psi_S(0)\rangle = |\psi_S(0)\rangle \tag{6.315}$$

where $|\psi_H\rangle$ is the time-independent basis in the Heisenberg representation. While the basis are now fixed in time, the time dependence is shifted to the operators $\hat{A}_H = \hat{A}_H(t)$. If $\hat{A}_S$ is a fixed operator in the Schrödinger representation, using eq. 6.315 we can express the same operator in Heisenberg representation by calculating the expectation value:

$$\begin{aligned}\langle\hat{A}_S\rangle &= \langle\psi_S(t)|\hat{A}_S|\psi_S(t)\rangle \\ &= \langle\psi_S(0)|e^{i\frac{\hat{H}}{\hbar}t}\hat{A}_S\,e^{-i\frac{\hat{H}}{\hbar}t}|\psi_S(0)\rangle \\ &= \langle\psi_H|\hat{A}_H|\psi_H\rangle\end{aligned} \tag{6.316}$$

where the operator $\hat{A}$ in the Heisenberg representation is related to its Schrödinger representation by:

$$\hat{A}_H(t) = e^{i\frac{\hat{H}}{\hbar}t}\hat{A}_S\,e^{-i\frac{\hat{H}}{\hbar}t}. \tag{6.317}$$

We can calculate explicitly the time evolution of the Heisenberg operator by cal-culating the time derivative:

$$\begin{aligned}\frac{d\hat{A}}{dt} &= \frac{i}{\hbar}\hat{H}e^{i\frac{\hat{H}}{\hbar}t}\hat{A}e^{-i\frac{\hat{H}}{\hbar}t} - \frac{i}{\hbar}e^{i\frac{\hat{H}}{\hbar}t}\hat{A}\hat{H}e^{-i\frac{\hat{H}}{\hbar}t} + \frac{\partial\hat{A}}{\partial t} \\ &= \frac{i}{\hbar}[\hat{H},\hat{A}] + \frac{\partial\hat{A}}{\partial t}.\end{aligned} \tag{6.318}$$

Eq. 6.318 is called **Heisenberg equation** and governs the time dependence of the operators in the Heisenberg representation. Heisenberg equation 6.318 is equivalent to the Poisson brackets equation 1.85 of classical mechanics.

It was realized by P.A.M. Dirac[33] that it is possible to make a transition from classical mechanics to quantum mechanics by reinterpreting all the classical variables as operators with the important addition of imposing a commutation relation on the fundamental operators. The Poisson brackets $\{q, p\} = 1$ of classical mechanics then become the commutator $[q, p] = i\hbar$ of quantum mechanics.

Dirac has also introduced the so called **interaction picture** where both the operators and the basis depend on time. The interaction picture is useful when the Hamiltonian of many particles can be written as the sum of two terms $H = H_0 + H_{int}$ where H_0 is the Hamiltonian for the free particles plus an interaction term H_{int}.

6.6   WHY THE IMAGINARY NUMBER i?

The mathematics of quantum mechanics that we have described in this book makes heavy usage of complex numbers. On the other hand, all observable quantities like, for example position, momentum, energy, etc., are described by real numbers. We have seen that both Schrödinger and Heisenberg equations describing the time evolution of a quantum system contain the imaginary number "i". In fact, operators representing observables are required to be hermitian, which possess the property of having real eigenvalues representing the range of possible outcome of a measurement.

It is interesting to try to investigate the reasons for the appearance of complex numbers in quantum mechanics. If the reader were to poll a statistically significant number of physicists to why quantum mechanics requires complex numbers, the majority will probably answer "because it works!". This answer is undoubtedly acceptable because, as of today, quantum mechanics seems to be the best theoretical framework to interpret a large variety of experimental data. In other words, we do not know of a single experimental evidence so far against quantum mechanics[34].

Let's start with pointing out that, in the Heisenberg representation, the physics is described by commutators between hermitian operators. Let's call the operator C as the commutator of two hermitian operators A and B, i.e. C=[A,B]. Since A and B are hermitian, the operator C must also be hermitian if it is to represent an observable.

---

[33]Paul Adrien Maurice Dirac (1902-1984) was an English physicist who made fundamental contributions to the early quantum mechanics and, later, to quantum electrodynamics. Through his Dirac equation he correctly described the fermions and predicted the existence of antiparticles. He was awarded the Nobel prize in physics in 1933.

[34]We do know that quantum mechanics and general relativity cannot be both valid at small scales in the presence of strong gravitational fields, i.e. high energy scales. As an example of such incompatibility, consider Heisenberg uncertainty principle stating that we cannot measure simultaneously momentum and position. General relativity tells us that momentum alters the flat space-time which, in turn, determines how position is measured. This kind of feedback is not yet described by both quantum mechanics and general relativity.

Using the property $(AB)^\dagger = B^\dagger A^\dagger$, we have:

$$C^\dagger = [A,B]^\dagger = (AB)^\dagger - (BA)^\dagger = [B^\dagger, A^\dagger] = [B,A] = -[A,B] = -C \qquad (6.319)$$

i.e. the operator C is anti-hermitian. The *only* way to make it hermitian would be to multiply it by i. In fact, if we define $C = i[A,B]$ it follows immediately that C is hermitian. Multiplication by $\hbar$ is then needed in order to insure the correct physical dimensions like Heisenberg position-momentum commutator $[q,p] = qp - pq = i\hbar$.

Having established that operators can be complex, we now turn our attention to the wavefunction. Can we have quantum mechanics with real wavefunctions and complex operators? After all, classical electrodynamics allows electromagnetic waves written as:

$$E_x = E_0 \cos(kz - \omega t) \qquad (6.320)$$

describing a plane wave propagating along the x direction.

We know that the classical wave 6.320 can be decomposed into two components characterized by positive and negative frequency. In fact, using Euler identity on eq. 6.320 we have that:

$$E_x = \frac{E_0}{2} \left(e^{i(kx - \omega t)} + e^{-i(kx - \omega t)} \right). \qquad (6.321)$$

Eq. 6.321 tells us that any **real** sinusoidal signal can be decomposed into the sum of a positive and a negative frequency component. This result can be extended to any real oscillating observable. In fact, given a real sinusoidal signal $f(t)$, its Fourier transform is:

$$\mathscr{F}(\omega) = \int_{-\infty}^{\infty} f(t) e^{-i2\pi\omega t} \qquad (6.322)$$

where we will always have:

$$|\mathscr{F}(\omega)| = |\mathscr{F}(-\omega)|. \qquad (6.323)$$

Eq. 6.321 can be interpreted as follows: any real sinusoidal signal can be decomposed into the sum of two complex sinusoids each representing, respectively, the negative $(i\omega t)$ and positive $(-i\omega t)$ frequency components. The specific combination of the two complex sinusoids gives as a result a real sinusoid. Complex sinusoids are constantly used in physics and engineering instead of real sine or cosine providing that at the end we consider only the real part and discard the imaginary part. The combination 6.321 insure that the imaginary parts cancel exactly out. Complex sinusoid s have also the convenient characteristics of having a constant modulus which, as we will see later, is quite important in quantum mechanics.

We have seen previously in this book that there is ample experimental evidence that massive particles like electrons, for example, need to be associated with a wave with wavelength $\lambda = \frac{h}{p}$ in order to explain interference phenomena. The question is: can we use a real wave like 6.320 to describe the wave associated with a massive particle? If the particle has a non relativistic energy equal to $\frac{p^2}{2m}$, then it must have positive energy is its mass is positive. In addition, if the energy is positive then, due to Planck formula $E = h\nu$, the wavefunction must contain only the positive energy part of eq. 6.321. This means that the wavefunction must be of the form:

$$\psi \propto e^{i(kx - \omega t)} \tag{6.324}$$

which means that the wavefunction has to be complex.

This argument is reinforced if [40] [21] we consider Born's interpretation of the wavefunction together with Heisenberg uncertainty principle. According to Heisenberg, we have seen that if we consider a particle with a well-defined momentum p_x along the x-axis, we cannot make any prediction about its x-position. Following de Broglie, and neglecting the time dependency, let's try to express the associated wavefunction in terms of wavelength as a real function of the form:

$$\psi(x) = A \cos \frac{2\pi x}{\lambda} = A \cos \frac{p_x x}{\hbar} \tag{6.325}$$

where we have used de Broglie relation in the last equality. Let's now assume that we know exactly the value of p_x. In this case we must have complete uncertainty in the position of the particle. If we are to use Born's interpretation of the wavefunction 6.325, its square modulus should reflect the fact that we have no information about its position, i.e. its square modulus should be a constant. In fact, a constant square modulus tells us that the particle has equal probability to be anywhere along the x-axis thus satisfying Heisenberg uncertainty principle. However, the square modulus of 6.325 is:

$$|\psi(x)|^2 = |A|^2 \cos^2 \frac{p_x x}{\hbar} \tag{6.326}$$

which is an oscillating function with periodic maxima and minima. This means that we know that the particle can be with higher probability at the locations of the maxima of 6.326 or lower probability at the locations of the minima. This violates the uncertainty principle.

On the other hand, we can construct an oscillating wavefunction for the particle of exactly known momentum p_x by assuming:

$$\psi_c(x) = A e^{\frac{i p_x x}{\hbar}}. \tag{6.327}$$

This wavefunction has an oscillatory behavior capable of taking into account interference and, at the same time, has a constant modulus given by:

$$|\psi_c(x)|^2 = \psi^*(x)\,\psi(x) = A^* A\; e^{-\frac{i p_x x}{\hbar}} e^{\frac{i p_x x}{\hbar}} = |A|^2 \tag{6.328}$$

which shows that we can have an oscillatory wavefunction with constant modulus providing that we allow the usage of complex functions.

We might try another attempt at formulating quantum mechanics using classical probabilities which are real numbers. In this particular formulation we can try to write probability vector and operators as stochastic matrices, i.e. square matrices whose elements are real numbers expressing probabilities between 0 and 1.

Following [21], let's consider a simple two-states system described by a column vector:

$$p_0 = \begin{pmatrix} 0.3 \\ 0.7 \end{pmatrix}. \tag{6.329}$$

For example, the state vector 6.329 could describe a spin one-half system, We can then interpret it as describing a system that has 30% probability of being in spin-up state and 70% probability of being in spin-down state.

An example of a transition from the state p_0 to a new state p_1 can be represented by the following stochastic matrix T:

$$T = \begin{pmatrix} 0.2 & 0.4 \\ 0.8 & 0.6 \end{pmatrix}. \tag{6.330}$$

Applying the matrix T to the state p generates a new state p_1 given by:

$$p_1 = T p_0 = \begin{pmatrix} 0.2 & 0.4 \\ 0.8 & 0.6 \end{pmatrix} \begin{pmatrix} 0.3 \\ 0.7 \end{pmatrix} = \begin{pmatrix} 0.34 \\ 0.66 \end{pmatrix}. \tag{6.331}$$

Notice that T is well normalized to represent a stochastic matrix having all the columns adding up to 1.

The transition 6.331 starts from the vector p_0 and ends up with the vector p_1 representing a new spin state with probability of spin up equal to 34% and probability spin down equal to 66%. A sanity check on the final states verifies that the probabilities still add up to 1.

If we now require that the matrix T takes the state p_0 at time t_0 and produces the state p_1 at time $t_1 > t_0$, we are describing a time evolution of the system. In this case we must impose that time evolution transitions are continuous or, in other words, that we can always decompose the transition $T(t_0, t_1)$ into two successive transitions $T(t_0, t_i)$ and $T(t_i, t_1)$ such that $T(t_0, t_1) = T(t_0, t_i) T(t_i, t_1)$ where t_i is an intermediate time t_i. More in general, when we consider time evolution in quantum mechanics we assume that processes can be subdivided into smaller and smaller time steps. If each step is described by a time evolution operator and assuming $n + 1$ equal time steps δt we must have that:

$$T(t_0, t_1) = T(t_0, t_0 + \delta t) T(t_0 + \delta t, t_0 + 2\delta t)...T(t_0 + n\delta t, t_1). \tag{6.332}$$

In the simple case of dividing the time interval into two steps, we have that:

$$T(t_0, t_1) = T(t_0, t_0 + \delta t) T(t_0 + \delta t, t_1) = T(t) T(t) = T^2(t) \tag{6.333}$$

where we called $T(t)$ the time evolution operator that evolves the system by a time interval δt. Continuity therefore implies the existence of the square root of the time evolution operator or, more in general, the n-th root.

Let's go back to our numerical example of eqs. 6.329, 6.330 and 6.331 and let's subdivide the time evolution matrix 6.330 into two successive matrices. This is achieved by taking its square root which can be lengthy[35]. Karam[21] gives the following expression for the square root of 6.330:

$$\sqrt{T} = \frac{1}{3} \begin{pmatrix} 1+i\frac{2}{\sqrt{5}} & 1-i\frac{1}{\sqrt{5}} \\ 2-i\frac{2}{\sqrt{5}} & 2+i\frac{1}{\sqrt{5}} \end{pmatrix} \tag{6.334}$$

showing that we obtain a matrix with imaginary elements. $\sqrt{T}$ cannot represent a stochastic matrix which is contradicting our initial assumptions. Therefore if we require continuity we are forced to use complex numbers.

---

[35]For a procedure to calculate the square root of a 2×2 square matrix see Levinger [24].

References

1. *CRC Handbook of Chemistry and Physics*, Edited by John Rumble, CRC Press, 2022.
2. V. I. Arnold. *Mathematical Methods of Classical Mechanics*. Springer-Verlag, 1989.
3. A. B. Arons and M. B. Peppard. Einstein's Proposal of the Photon Concept. *American Journal of Physics*, 33(5), 1965.
4. L. E. Ballantine. *Quantum mechanics: a modern development*. World Scientific Publishing Co. Pte. Ltd., 1998.
5. N. Bohr. The structure of the atom. *Nobel lecture*, 1922.
6. M. Born. Zur quantenmechanik der stossvorgange. *Zeitschrift für Physik*, 37, 1926.
7. F. W. Byron and R. W. Fuller. *Mathematics of classical and quantum physics*. Dover Publications, Inc., 1992.
8. R. Resnick and D. Halliday. Physics. 1980.
9. R. Das. Wavelength and frequency-dependent formulations of wien's displacement law. Journal of Chemical Education, 92, 2015.
10. C. Davisson and L. H. Germer. The scattering of electrons by a single crystal of nickel. *Nature*, 119(2998), 1927.
11. L. de Broglie. Recherches sur la theorie des quanta (researches on the quantum theory). Annales de Physique, 3, 1925.
12. P. A. M. Dirac. *The Principles of Quantum Mechanics*. Oxford University Press, 1958.
13. P. Ehrenfest and H. Kamerlingh Onnes. Simplified deduction of the formula from the theory of combinations which planck uses as the basis of his radiation theory. Philosophical Magazine, 29(170), 1914.
14. A. Einstein. Planck's theory of radiation and the theory of specific heat. *Annalen der Physik*, 22, 1907.
15. R. P. Feynman. Space-time approach to non-relativistic quantum mechanics. *Reviews of Modern Physics*, 20(2), 1948.
16. R. P. Feynman. *The Feynman lectures on physics*. Addison-Wesley, 1977.
17. G. Gamow. *Thirty years that shook physics*. Doubleday and Co. Inc., 1966.
18. W. Gerlach and O. Stern. Der experimentelle nachweis der rich- tungsquantelung im magnetfeld. *Zeitschrift für Physics*, 9, 1922.
19. D. J. Griffiths. *Introduction to quantum mechanics*. Prentice Hall, Inc., 1995.
20. W. Heisenberg. Über den anschaulichen Inhalt der quantentheoretischen Kinematik und Mechanik. *Zeitschrift für Physik*, 43, 1927.
21. R. Karam. Why are complex numbers needed in quantum mechanics? Some answers for the introductory level. *American Journal of Physics*, 88(1):39–45, 2020.
22. L. Kelvin. Nineteenth century clouds over the dynamical theory of heat and light. *Philosophical Magazine and Journal of Science*, 2(7), 1901.
23. M.P. Hobson K.F. Riley and S.J Bence. *Mathematical methods for physics and engineering*. Cambridge University Press, 2002.
24. B. W. Levinger. The square root of a 2×2 matrix. *Mathematics Magazine*, 53(4):222–224, 1980.
25. M. S. Longair. *Theoretical concepts in physics*. Cambridge University Press, 1994.
26. F. Mandl. *Statistical physics*. John Wiley and Sons, 1988.

27. S. A. C. McDowell. A simple derivation of the boltzmann distribution. *Journal of Chemical Education*, 76(10), 1999.

28. E. Merzbacher. *Quantum Mechanicxs*. John Wiley and Sons Inc., 1961.

29. R. A. Millikan. A direct determination of planck's h. Physical Review, 4, 1914.

30. R. A. Millikan. A direct photoelectric determination of planck's h. Physical Review, 7, 1916.

31. I. Newton. *Philosophiae Naturalis Principia Mathematica*. Edited by Edmond Halley, 1687.

32. H.C. Ohanian. What is spin? *American Journal of Physics*, 54, 1986.

33. O. Passon. Kelvin's clouds. *American Journal of Physics*, 89(11), 2021.

34. T. E. Phipps and J. B. Taylor. The Magnetic Moment of the Hydrogen Atom. Physical Review, 29, 1927.

35. L. Piccirillo. *Introduction to the maths and physics of the solar system*. CRC Press, 2020.

36. H. Rubens and F. Kurlbaum. On the heat radiation of long wave-length emitted by black bodies at different temperatures. *Astrophysical Journal*, 14, 1901.

37. E. Rutherford. Retardation of the α particle from radium in passing through matter. *Philosophical Magazine and Journal of Science*, 12, 1906.

38. L. I. Schiff. *Quantum Mechanics*. McGraw-Hill, 1955.

39. C. T. Sebens. How electrons spin. *Studies in History and Philosophy of Modern Physics*, 68, 2019.

40. R. Shankar. *Principles of Quantum Mechanics - Second edition*. Springer, 1994.

41. A. Small and K. S. Lam. Simple derivations of the hamilton–jacobi equation and the eikonal equation without the use of canonical transformations. *American Journal of Physics*, 79, 2011.

42. G. E. Uhlenbeck and S. Goudsmit. Ersetzung der hypothese vom unmechanischen zwang durch eine forderung bezüglich des inneren verhaltens jedes einzelnen elektrons. *Die Naturwissenschaften*, 13, 1925.

43. S. Mandelstam and W. Yourgrau. *Variational Principles in Dynamics and Quantum Theory*. Pitman Publishing Corporation, 1960.

44. B. R. Wheaton. Lenard and the photoelectric effect. *Historical Studies in the Physical Sciences*, 9, 1978.

45. W. Wien. On the division of energy in the emission-spectrum of a black body. *Philosophical Magazine*, 43, 1897.

Index

Printed in the United States
by Baker & Taylor Publisher Services